Springer Series in Language and Communication 5

Editor: W. J. M. Levelt

Springer Series in Language and Communication

Editor: W. J. M. Levelt

Volume 1 **Developing Grammars**
By W. Klein and N. Dittmar

Volume 2 **The Child's Conception of Language**
Editors: A. Sinclair, R. J. Jarvella, and W. J. M. Levelt

Volume 3 **The Logic of Linguistic Development in the Early Childhood**
By M. Miller

Volume 4 **Inferring from Language**
By L. G. M. Noordman

Volume 5 **Retrieval from Semantic Memory**
By W. Noordman-Vonk

Volume 6 **Semantics from Different Points of View**
Editors: R. Bäuerle, U. Egli, A. von Stechow

Volume 7 **Lectures on Language Performance**
By Ch. E. Osgood

Wietske Noordman-Vonk

Retrieval from Semantic Memory

With a Foreword by John C. Marshall

With 10 Figures

Springer-Verlag Berlin Heidelberg New York 1979

Dr. Wietske Noordman-Vonk

Department of Psychology, University of Groningen
The Netherlands

Temporary address:

Department of Psychology, Carnegie-Mellon University
Pittsburgh, PA 15213, USA

Series Editor:

Professor Dr. Willem J. M. Levelt

Max-Planck-Gesellschaft zur Förderung der Wissenschaften e.V.
Projektgruppe für Psycholinguistik
Berg en Dalseweg 79, Nijmegen, The Netherlands

ISBN-13:978-3-642-67217-0 e-ISBN-13:978-3-642-67215-6
DOI: 10.1007/978-3-642-67215-6

Library of Congress Cataloging in Publication Data. Noordman-Vonk, Wietske, 1939– Retrieval from semantic memory. (Springer series in language and communication; v. 5). Bibliography: p. Includes indexes. 1. Memory. 2. Semantics. I. Title. II. Series. BF371.N56 153.1'3 78-31354.

Softcover reprint of the hardcover 1st edition 1979

2153/3130-543210

Foreword

The area of concern to Dr. Wietske Noordman-Vonk has been variously seen as an aspect of long-term memory [F.1], secondary memory [F.2], memory without record [F.3], and semantic memory [F.4], the latter term being the one preferred by Dr. Noordman-Vonk herself. This proliferation of terminology is not an entirely trivial matter, for although the expressions clearly overlap in range, they do draw attention to different features of the phenomena under consideration.

The work reported here is concerned with the form of representation and manipulation of our knowledge that, for example, a dog is an animal, or that mothers and daughters are parents and children. To put it more generally, the experiments attempt to elucidate the psychological processes involved in the semantics of class-inclusion and, most importantly, to extend the explanatory principles there invoked to a new domain, that of kinship relations. Clearly, the connections between "ant" and "insect", or "flower" and "plant" have been known to us - as adults - for some considerable period of time; in the absence of brain injury or degeneration we are unlikely to "forget" that fathers and sons are kin of the same sex. We may therefore - pretheoretically - distinguish between retrieval of such knowledge and retrieval of a rapidly fading sequence of random numbers that we are asked to recall after a single presentation. It is in this sense that the current work is concerned with long-term and not short-term memory.

William James, on the other hand, distinguishes between primary memory, which is that part of the present during which incoming signals are analysed, and secondary memory which includes stored knowledge of facts. Given our everyday system of belief, we may readily concede that the proposition "A dog is an animal" represents a (true) fact. We may further agree with James that we do not frequently "think this fact"...... unless prompted by an experimental psycholinguist. However, when we do so, it is with full "...consciousness that we have thought or experienced it before" [F.2]. The idea does not suddenly appear to us as a novel insight into theoretical biology!

Bruner's terminology again changes the emphasis. He writes of *memory with record* where a specific event - perhaps with a time tag - can be recovered, as in my recall of the delicious lamb that my wife cooked for my last birthday. Recollection of such unique events is then distinguished from *memory without record,* "... where encounters are converted into some process that changes the nature of an organism, changes his skills, or changes the rules by which he operates but which are virtually inaccessible in memory as specific encounters" [F.3]. The child's developing command over the structure of lexicon and syntax, and the implications of this for the growth of reasoning are *prima facie* changes in the nature of the organism.

Finally, we turn to the semantic aspect. In order to count as a mature speaker of Dutch, one must know, at least tacitly, that "Een kanarie is een vogel". (I do not suggest this is the *sole* knowledge required!) One cannot, in this sense, know the meaning of "canary" without knowing that canaries are birds. Such facts are part and parcel of the semantic structure of the language.

How, then, are we to account for the vast store of permanent, skilled, highly available knowledge that we draw upon in talking and thinking? What form of organization allows us such ready and (relatively) error-free access to the pattern of interconnection of our vocabulary and its deployment in the expression of propositions?

As always in psychology, the first step towards an answer is to see that there is a genuine question here. As WITTGENSTEIN [F.5] notes, we are dealing with "...observations which no one has doubted"; they may indeed "escape remark" just because "they are always before our eyes".

But consider: An educated adult will have a vocabulary in the order of 75,000 words; these words will enter into a (potentially) infinite number of true propositions and a similarly infinite number of false propositions. Now, what Dr. Noordman-Vonk's experiments demonstrate is that in a little over a second, subjects can show that they know, for example, that "oak" and "beech" are exemplars of the superordinate category "tree"; in a time which is only slightly longer, the subjects can show that they know that, for example, "oak" and "ant" are not *both* exemplars of any single category drawn from the set "tree", "flower", "bird", and "insect". These reaction times include, of course, the time taken to read the stimulus items and the time it takes to program and execute a motor response; the value we are interested in, namely semantic decision time, must therefore be very considerably less than a second.

What form must our semantic knowledge take in order to allow such rapid retrieval and decision? Aristotle [F.6] proposed that we think of the problem by analogy with a library. We are invited to consider words as volumes in a library; these volumes (= words) are arranged on shelves according to a metric of semantic distance, and they are furthermore hierarchically stacked so that volumes on general topics are to be found on "higher" shelves than volumes containing more specific information. Within such a system, Aristotle can talk of remembering being "... a kind of search". It is this spatial metaphor which allows experimentalists to conceive of using *time* as a measure of *structure*. The model suggests that we can measure the (hidden) distance from "oak" to "tree" by starting from "oak" and seeing how long it takes to arrive at "tree".

Scholars have sometimes affected to take the model quite literally. Thus CARUS [F.7] wrote: "If we would take out of .... a brain the living substance without destroying the membranes in which the cells have enveloped themselves, it would afford an aspect of divisions and subdivisions not unlike that of the departments, shelves, and pigeon holes of a library from which the books are removed, and we would have an anatomical representation of a system of formal thought".

As Dr. Noordman-Vonk points out, the usual modern notation for the library theory takes the form of a *network model*, "a system of nodes and pointers". An alternative notation that she discusses - found in so-called *set theoretic models* - represents concepts as collections (perhaps ordered) of semantic features. This notation also has its analogue in earlier systems, for it is essentially a card-index system in which the relationship between books (= words) is captured by the number of key-terms that they hold in common; CARUS [F.7] introduced the idea in these words: "Let us now suppose that the chief librarian of the library of our brains for the sake of arranging a catalogue takes an inventory of all the books...".

Although one might have supposed that network and feature set models are instances of different theories, it would seem that over a considerable range of data the models are (or can be made to be) isomorphic [F.8]. As Dr. Noordman-Vonk remarks, the distinction is probably not fundamental to current attempts at explication of the word-store.

I have already mentioned that CARUS (and indeed many others) could take the spatial component of both the library and the library-index models in a straightforwardly anatomical sense. It is, of course, more conventional nowadays to think of space metaphorically. Semantic *distance* is regarded as a "psychological" scale of measurement, not a "physical" one; likewise, the *length* of a feature-list.

No commitment is made as to the nature of the underlying neuronal realization of the system. No doubt this is the correct attitude, but whatever "real" mechanism we believe to produce the results of semantic verification tasks, that mechanism must somehow meet the conditions laid down by the pattern of reaction times derived from the spatial metaphor.

Consistent with the English-language literature, Dr. Noordman-Vonk's experiments show that there are three crucial results that any model must explain: The "nearness" effect, whereby judgments that, for example, canaries are birds can be made more quickly than judgments that they are animals; the "reverse nearness" effect, whereby judgments that canaries are *not* fishes takes longer than judgments that they are not flowers; the "same category" effect whereby judgments of *true* (or *same category* for two stimuli) are faster than judgments of *false* (or *different category*). These "semantic distance effects", and their interaction with response category, constitute the primary data to which theory must be responsive.

It is easy to produce an information-processing model that accounts for any one of these effects; it is fairly easy to construct hypotheses that account for two of them, and it is very difficult to make a well-motivated model for all three. It is this last goal that Dr. Noordman-Vonk has set herself. She has succeeded, for her process-models give an extremely good fit to the data within two quite distinct semantic domains.

To interpret her reaction-time studies of class-inclusion, Dr. Noordman-Vonk proposes two explanatory principles. First, she suggests that positive ("confirming") evidence will be sought for the truth of the experimental propositions, and that only if such evidence is not forthcoming will a falsification routine be employed. Second, she suggests that the *particular* order of deployment of stored semantic information within these two stages depends conjointly upon the demands of the task and the criterion set by the experimenter. The generality of these principles is then demonstrated by showing how they can be deployed in the interpretation of another semantic domain (kinship terms) and another task (judgments of semantic similarity between concepts).

One might complain that the specific models that are constructed to instantiate these principles are rather complicated; many boxes, many substages are involved. But this is no criticism unless one can produce a more parsimonious account that covers the data at least as well.

One might be surprised that the models take the form of strictly serial - first try this, and then that - operations. Would it not be simpler to evaluate semantic relationships in a parallel fashion? Perhaps so, but that system would have to be equivalent on some level to the models described here [F.9].

Finally, one might wonder whether the machinery set up to perform semantic verification tasks is specific to the process of explicit decision-making. If so, then the studies reported here fall within the domain of theories of reasoning. They are, if you like, concerned with meta-control over the internal structure of our vocabulary. This was how CARUS regarded the cataloguing activities of his cerebral librarian: "This reference room in our brain is called logical ability ... and its establishment marks another important step in the development of reason; it is formal thought. It is the beginning of scientific thought by the help of which we gain information about the methodical arrangement of our conceptions". Some of the connections between semantic memory and the inferential processes involved in integrating information derived from sets of sentences and from "real world" knowledge are explored in the companion volume of Dr. Leo Noordman [F.10].

Dr. Noordman-Vonk seems to have such an extended paradigm in mind when she discusses the crucial "role of the criterion in actualizing the structure of the domain". If this is the right way of looking at the results, we could now begin to ask such questions as what is the relationship between *explicit* semantic decision-making and the "unconscious" processes involved in the lexical selection and recognition component of language production and perception models. Such questions are obviously bound up with arguments concerning the existence - or otherwise - of a distinct level of semantic representation in language comprehension [F.11]; resolution of this issue will in turn affect the interpretation of on-line semantic effects in the processing of continuous speech [F.12]. Eventually we might hope to forge closer links between the psychology of reasoning and the psychology of language: "Logic does not create order and system in our brain, but it makes us conscious of the order that grew naturally in our mind" [F.7]. This line of argument was recently investigated by BRAINE [F.13] in his studies of the relationship between the inaccessible inference schemata that underlie reasoning and our conscious knowledge of the products or results of inferential processes.

Let me conclude by mentioning one further link: Dr. Noordman-Vonk's studies clearly continue the pattern of method and theory first laid down by DONDERS [F.14] in The Netherlands, and by WUNDT [F.15] and his students in Germany. How good it is to see the re-establishment of this vital tradition in European psychology.

Nijmegen, November 1978 John C. Marshall

Acknowledgments

I am greatly indebted to the series editor Dr. W.J.M. Levelt for his continued encouragement, discussions, helpful suggestions, and for his comments on the manuscript. Thanks are also due to Dr. J.A. Michon who has read and commented on the whole manuscript. I wish to thank the staff of the Institute for Experimental Psychology of Groningen University for the construction of the equipment, many students and research assistants for their help in conducting experiments, and Ms. E. Maarschalkerweerd for typing carefully the manuscript.

This research was supported in part by a grant from the Netherlands Organization for the Advancement of Pure Research (Z.W.O.) to the author and to Dr. L.G.M. Noordman, and at an earlier stage by a grant from the Niels Stensen Stichting.

Contents

Chapter 1 Introduction 1

Chapter 2 Statistical Analysis of the Reaction Time Data 10

Part I A Model for Verifying Semantic Relations

Chapter 3 The Selection of the Material from the Hierarchical Domain .. 17

3.1 Experiment 1: The Production of Exemplars 18
3.2 Experiment 2: The Grouping of the Exemplars 21
3.3 Experiment 3: The Categorization of the Exemplars 23
3.4 Semantic Relations Between Two Exemplars 24
3.5 Experiment 4: The Categorization of Pairs of Exemplars 26

Chapter 4 The Verification of Hierarchical Relations 29

4.1 Experiment 1 29
4.1.1 Method 30
4.1.2 Results 32
4.1.3 Discussion 33
4.2 Experiment 2 36
4.2.1 Method 36
4.2.2 Results 37
4.2.3 Discussion 38
4.3 Experiment 3 39
4.3.1 Method 40
4.3.2 Results and Discussion 41
4.4 Experiment 4 42
4.4.1 Method 45
4.4.2 Results 45
4.4.3 Discussion 46
4.5 Conclusion 47

Part II Generalization to Another Task

Chapter 5 Judging the Semantic Similarity in a Hierarchical Domain 51

5.1 Method 53
5.2 Results 54
5.3 Discussion 60

Part III Generalization to Another Semantic Domain

Chapter 6 The Kinship Domain 63

6.1 Experiment 1: Judging the Semantic Similarity Between Kinship Terms 65
6.2 Experiment 2: The Production of Kinship Terms 65

Chapter 7 The Verification of Kinship Relations 69

7.1 Experiment 1 70
7.1.1 Method 71
7.1.2 Results 72
7.1.3 Discussion 73
7.2 Experiment 2 75
7.2.1 Method 76
7.2.2 Results 77
7.2.3 Discussion 78
7.3 General Discussion and Conclusion 81

Chapter 8 Summary and Conclusion 83

8.1 Summary 83
8.2 Conclusion 86

References 89
Author Index 93
Subject Index 95

Chapter 1 Introduction

Until recently, memory has been studied in precisely those aspects in which memory fails: the *trials* to store information and the *errors* in recalling information. Memory has been conceived of as learned responses. Studies on memory were actually concerned with learning and with recall and recognition of the learned material [1.1]. Accordingly, memory was often discussed under the heading of learning [1.2] or even under rote learning [1.3]. Memory is being studied at those points at which there is either not yet or no longer any retention. This is also clear from what is measured in the experiments: the dependent variable is the number of errors in recognition and recall ("no longer") or the number of trials in learning ("not yet").

Hardly any attention has been lavished upon another aspect of memory processes, viz., the nature of the stored material and its structure. Originally, the studies were mainly concerned with nonsense syllables in order to exclude the influence of the material as well as the dependency of the memory activities on the material. In the course of time, the nature of the material became relevant in the sense that properties of the material, such as meaningfulness and discriminability, are factors that are being studied experimentally. These factors appeared to have an influence anyway in experiments with nonsense material.

The kind of studies discussed so far concern what, since TULVING [F.4], commonly has been called episodic memory. TULVING made a distinction between semantic and episodic memory. Semantic memory contains knowledge that has become rather independent of the time and place in which it was acquired. Episodic memory refers to the specific experiences that occurred in time and space; information is stored with a tag for time and/or place of the acquisition. This distinction is certainly useful with respect to the method of research.

Semantic memory may be described as memory for meaningful information. It is the organized store of knowledge that a person possesses, about language and its use as well as about the world [Ref.1.4, p.284] (but see also [1.5]). Although not necessarily, the information that is being studied is mainly verbal information: words, sentences, text. While earlier research was mainly concerned with the acquisition of information and forgetting of information, semantic memory deals with actualization of the stored information. The central question is what information is stored and which processes underly the use of that knowledge. Another difference is related to the previous. As pointed out in [1.6], studies in semantic memory do not pay attention to learning or forgetting. Learning and forgetting are separated from each other now: they encompass a life span and not only an experiment. The acquisition of semantic memory, however, should constitute an important aspect of semantic memory research. Research on semantic memory differs also from the earlier memory research with respect to the experimental paradigms. Errorless responses would have made the earlier research impossible, because no measures on the dependent variable would have been obtained. On the other hand, most experimental paradigms in semantic memory at the present time require nearly errorless responses as a condition for the research: the dependent variable is the reaction time (RT), the time it takes a subject to perform a task.

The objective of these studies is to describe the processes within the same theoretical approach of information processing in which other complex human behavior such as perception, concept learning, and problem-solving can be characterized. By manipulating the information to be processed in specific ways and by relating this to the RTs, the processes are analyzed in a number of successive mental operations. This method was first used by by DONDERS [F.14] and more recently was worked out especially by STERNBERG [1.7].

While the information processing approach has been influential with respect to the functional analysis of semantic memory, linguistic studies have been influential with respect to the structural approach in semantic memory research. For generations, linguists have paid attention to the structure of the lexicon and, more generally, to semantic issues. Psychologists have been studying word meanings by collecting judgments of similarity between words. These judgments are analyzed by means of cluster analyses and scaling methods [1.8-12]. From the similarity judgments one infers to the subjective organization of the lexicon. This organization is frequently described in terms of bipolar features that are hierarchically organized.

The domain of semantic memory research is usually conceived of as rather restricted: the research is directed to the development of detailed, local models. The work of COLLINS and QUILLIAN [1.13], SMITH et al. [1.14] and GLASS and HOLYOAK [1.15] is relevant at this point. Evidently, large-scale models such as the models of ANDERSON and BOWER [1.16], KINTSCH [1.17], and NORMAN and RUMELHART [1.18] are also concerned with semantic memory. They deal with the conceptual representation of knowledge within the broader framework of cognitive functioning and language processing, and do not limit themselves to the semantic part of Tulving' s distinction. ANDERSON and BOWER, for example, explained all knowledge on the basis of particular experiences.

The present study, however, is in the area of the more detailed models. No review on that domain is given, because some recent review articles are available [1.5,19,20]. But a characterization of the main experimental results and of the models is presented in order to outline the framework of the present study.

The experimental work has almost exclusively been concerned with the relation of inclusion, as in *a dog is an animal*. Thus, the domains that have been studied are hierarchically organized domains.

The tasks that generally have been used in the study of semantic memory are simple verification tasks. One can distinguish two types of tasks. In one type, subjects are required to judge whether a relation of inclusion between two concepts exists (true/false judgment). The concepts are presented in a sentence with the frame: a S (subject) is a P (predicate), e.g., *a collie is a dog, an oak is an animal* [1.13,21], or as two concepts, in which case it is made clear which concept has to be treated as the category and which one as the exemplar, e.g., is *collie* an exemplar of the category *animal* [1.22,23]. Some variations of the sentence frame are used as well, e.g., *all S are P, some S are P* [1.24,25]. The main independent variable is the nature of the semantic relation between the concepts. The RT for judging the sentence *a canary is a bird* is compared with the RT for *a canary is an animal* (true sentences) and the RT for judging *a canary is a sparrow* with the RT for *a canary is a fish* (false sentences). It should be mentioned that the sentences in some studies express not only inclusion relations but also property relations (e.g., *a canary has feathers*) in the same hierarchical domain [1.13].

In the second type of task, two concepts are presented, e.g., *oak beech*, and the subject has to judge whether both concepts are exemplars of a prespecified category, e.g., *tree*. In this task, concepts are judged as "same" or "different" with respect to prespecified categories [1.26-29]. The main independent variable is the nature of the relation between the two concepts or between each of the concepts and the prespecified categories. For example, the RTs for the following items, to be judged with respect to the prespecified category *bird*, are compared: *sparrow robin*, *sparrow ant*, *sparrow oak*.

Fundamentally the tasks are comparable. The RT appears to be a function of the semantic similarity between the concepts in both tasks. The RT for the answer "true" or "same" turns out to be shorter the more similar the words are. This will be called the semantic distance effect for the true or same items. The RT for the answer "false" or "different" is longer the more similar the words are. This will be called the semantic distance effect for the false or different items. The RT for *a canary is a bird* is shorter than the RT for *a canary is an animal*; the RT for *a canary is a sparrow* is longer than for *a canary is a fish*. Likewise, when judging concepts with respect to the category *animal*, the RT for th word pair *sparrow robin* is shorter than the RT for *sparrow ant*; and when judging with respect to the category *bird*, the RT for *sparrow ant* is longer than the RT for *sparrow oak*. Finally, one can frequently observe that the RT for the answer "true" or "same" is shorter than for the answer "false" or "different" (cf [1.30]).

The fact that the results in both tasks are similar suggests that the data might be accounted for by a single model. The question is what would such a model look like? The models that have been proposed can be distinguished in network models and set-theoretic models, a distinction formulated by SMITH et al. [1.19]. A network is a system of nodes and pointers; the nodes represent concepts, the pointers represent the relations between the concepts. The retrieval process is considered to be the most important process in verifying semantic relations. When verifying the sentence *a canary is a bird*, the network is entered at the nodes *canary* and *bird* and a path is searched between these nodes.

In set-theoretic models, concepts are represented as sets of features [1.14,24,26,27]. Sampling and decision processes are the most important processes in verifying semantic relations. The decisions are based on the information that is sampled from the relevant sets of features. Some set-theoretic models require that the similarity between concepts, indicated by the number of same elements, exceeds a certain threshold in order for an answer to be given [1.19,27]. According to the other models, the process is a search to find out whether each attribute of a category is an attribute of the exemplar [1.24] or whether the name of the exemplar is an element of the set that is stored with the category [1.31].

How do these models account for the experimental data? In the most parsimonious version of a network model, semantic similarity between two concepts is represented as the distance between the concepts [1.13]. The distance is expressed by the number of intermediate nodes between the concepts. Similarity is inversely related to the distance. COLLINS and QUILLIAN proposed a model in which the retrieval process is essentially a travel in semantic space: the length of the retrieval process is proportional to the semantic distance between the two concepts of the sentence. This model, however, does not account for the data in false sentences. In their opinion these could be explained by the "search and destroy" hypothesis. According to this hypothesis a subject tries to find a path which connects the subject and the predicate of the sentence that has to be verified. When such a path has been found, he has to check whether or not the path corresponds to the relation that is stated in the sentence. The number of paths to check is inversely related to the semantic distance. Therefore, the hypothesis accounts for the data on the false sentences. Because, however, the hypothesis cannot account for the true sentences as well, COLLINS and QUILLIAN in fact proposed two models for retrieval processes, one for the true sentences and one for the false sentences. As was mentioned earlier, COLLINS and QUILLIAN dealt not only with exemplars and categories but also with properties of those concepts. This, however, raised some difficulties for their model, too [1.32] (but see also [1.20]).

COLLINS and LOFTUS [1.20] elaborated upon this model more fully, especially with respect to the process of evaluation of the paths between the concepts. The paths differ in "criteriality". They were given differential weights indicating how essential each path is to the meaning of a concept (cf [1.33]), and they were given differential accessibility (i.e., strength or travel time) by COLLINS and QUILLIAN [1.34]. In [1.20], finally, paths were assumed to differ in weights for reaching a criterion for a positive or negative decision, even to such an extent that one path (e.g., superordinate link) can be sufficient. This holds true even for highly related concepts which are connected by many paths. In this way, the "search and destroy" hypothesis was extended to account for the data on true sentences. However, it is not quite clear how the weights of the paths are to be determined, and it is quite arbitrary to what extent the process is accounted for by differential weights of the paths and to what extent by the number of paths that are found between the concepts.

The most recent network model is the marker search model of GLASS and HOLYOAK [1.15]. Only category membership relations are represented. In this model, hierarchical connections can be cut short by direct pathways between markers that are hierarchically separated. Moreover, some paths from a marker are searched sooner than others (cf Collins and Quillian's criteriality in

[1.20]. Another important characteristic of the model is that the paths are labeled in such a way that a particular combination of the labels can give immediate information on the falsification of a sentence, e.g., in the form of a contradiction. The determinant of RTs is the order in which the paths are searched, for "true" as well as for "false" answers. The model was the first to propose a serious account for judgments on contradictory sentences. A question with respect to this account is, however, on which basis two paths have the same label so as to constitute the evidence for a "false" answer on, e.g., a contradictory sentence.

In a set-theoretic model, semantic similarity is defined in terms of the overlap of the sets of features of the concepts. The semantic similarity between *canary* and *animal* is smaller than between *canary* and *bird*, because the overlap, as expressed in the number of shared features, is smaller in the former case than in the latter. SCHAEFFER and WALLACE [1.27] proposed a feature set model in which the RT is a function of that overlap: the greater the overlap, the smaller the amount of information required for same judgments; the greater the overlap, the greater the amount of information required for different judgments. The information is obtained by a sampling process. However, the role of the categories with respect to which the words have to be judged is not specified in the sampling process. Furthermore, it is assumed that when one concept is identical to one of the prespecified categories, that concept is not analyzed into features. Therefore, no semantic distance effect should appear. This assumption is not only rather implausible but also disconfirmed by some of the data of SCHAEFFER and WALLACE [1.26].

The most elaborate set model is the feature-comparison model of SMITH et al. [1.19]. This model was constructed to account for the data on sentence verification experiments. It was also worked out for the results on word comparisons [1.29]. Word meanings are represented in terms of semantic features. A distinction is made between defining and characteristic features. For example, all birds have the defining features of the concept *bird*, e.g., "animate" and "feathered". Birds differ, however, to the extent in which they have the characteristic features of *bird*, e.g., "of a particular size" and "of a particular predacity". *Robin* and *bird* share more characteristic features than *chicken* and *bird* or *goose* and *bird*. This description fits in with the idea that category membership is a matter of degree rather than all or none [1.35]: *robin* is more typically a bird than *chicken* and *goose*. The feature-comparison model consists of two stages. The first stage, the similarity between the concepts, e.g., *sparrow* and *bird*, is assessed. The similarity is determined by the overlap of both the defining and the characteristic features. In case the overlap is greater than a certain threshold, the answer "true" is given. In case the overlap is smaller than another, lower, threshold, the answer "false" is given. In case the overlap has a value between these thresholds, the process enters a second stage in which the concepts are compared only with respect to their defining features. The answer "true" is given if the defining features of the category are also the defining features of the exemplar. Otherwise the answer "false" is given. This model explains why the RT for the true sentences is shorter the more similar the concepts are, and why the RT for false sentences is shorter the less similar the words are. The reason is that in both cases the second stage can be skipped, because in the first case the upper threshold is passed, and in the second case the lower threshold is not reached. It should be noted, however, that it is not clear on what basis the features are to be differentiated into defining and characteristic features (cf [Ref. 1.17, p.201]).

The distinction between network models and set-theoretical models is a convenient device to characterize the models as they have been described by the authors. This distinction is not, however, a fundamental distinction, as was pointed out specifically for the current semantic memory models by HOLLAN [F.8]. He argued that the available set-theoretic and network models are isomorphic.

The results that have commonly been obtained in research on semantic memory have not received equal attention. Most authors have paid attention to what were earlier called the semantic distance effects: the RT for true statements is shorter the more similar the concepts are; the RT for false statements is shorter the more different the concepts are. It is mainly to account for these data that the models were formulated. Surprisingly, almost no attention was paid to the result that the RT for true statements is shorter than for false statements (same/different effect).

Outside the field of semantic memory much work has been done on same/different judgments. In experiments of NICKERSON [1.36], for example, subjects had to decide whether the second of two visually presented consonants was the same as the first. RTs for the decision "same" were shorter than those for the decision "different". In other experiments of NICKERSON [1.37], subjects had to decide whether the second of two presented tones, that are the same or differ only in frequency, was the same as the first. The RT for "same" decisions was longer than that for "different" decisions. Same/different judgments were discussed by CALFEE [Ref. 1.38, p.262 ff.]. The results presented by CALFEE indicate that, if the stimuli to be compared can be coded, subjects do code the stimuli, and the RT for same judgments is at least as short as the RT for different judgments. If, however, the stimuli cannot be coded, as is the case with quantified variables, the RT for different judgments is shorter than for same judgments. This coding effect might be related to the categorical perception of speech sounds [1.39]. One would predict that if subjects can categorize speech sounds, the RT is shorter for same judgments than for different judgments whereas in cases where subjects cannot categorize speech sounds, but can differentiate between them, the results will be reversed. This might be investigated in comparisons of speech sounds that differ only with respect to vowel length. This line of research will not be pursued in this study.

The fact that the RT for true statements is shorter than for false statements might well be due to a general property of human cognition [Ref. 1.40, p.86]. The semantic memory models that were discussed do not pay attention to this finding. The model of SMITH et al., for example, could account for this finding by specifying in some way the values of the thresholds. Such an account, however, is hardly a theoretical explanation.

One of the objectives of the present study is to formulate a model that accounts for the same/different effect as well as for the semantic distance effects for true statements and for false statements.

As was stated earlier, research on semantic memory has primarily been concerned with hierachically organized material. In order for a model to have any validity, it has to be tested on a different semantic field as well. According to MILLER and JOHNSON-LAIRD [Ref. 1.41, p.237]: "Different linguists have used the term [semantic field] in different senses.... There seems to be a consensus that the term means something, but precisely what is open to question". A semantic field in the present study will denote a group of words that are closely related in meaning and that often are indicated by a general term [1.42]. According to CLARK and CLARK "semantic fields lie between two extremes, paradigms and taxonomies... most domains are a mixture of taxonomies and paradigms and natural categories like plant and animal names are mostly taxonomic" [Ref. 1.43, p.430]. A paradigmatic field is characterized by dimensions on which all words have a value [1.8]. A usual example is the field of kinship terms. The two kinds of organization were discussed by OSGOOD [1.9] in terms of a nested, ordered, independent feature hierarchy versus a replicated, unordered, independent feature matrix.

The second objective of the present study is to investigate the generalizability of the principles of the model for a field that is to be characterized as a paradigmatic field.

The present study consists of three parts. The first part deals with memory processes on hierarchically organized taxonomic material. The first experiments to be reported investigated the structure of the material. Based on the results of subsequent experiments, a model of mental processes underlying the verification of semantic relations was developed.

The second part concerns the generalization of the principles of the model to another task on the same material. The task as such does not appeal to the hierarchical organization of the material.

The third part deals with the generalization of the principles of the model to a semantic field with a paradigmatic organization: the field of kinship terms. The aim of the first experiments was to study the mental organization of the terms. In the subsequent experiments, verification tasks, comparable to the ones used in the first part, were used.

Chapter 2 Statistical Analysis of the Reaction Time Data [1]

As was pointed out in the previous chapter, the aim of the present study is to describe the temporal organization of cognitive processes. The tasks in this study require a subject to process semantic information. The usual experimental procedure is to manipulate the mental processes by varying the information to be processed. The dependent variable is the reaction time (RT), i.e., the time between the moment the information is presented and the moment the answer required by the task is given. By relating these times to the experimental manipulations, inferences are made with respect to the underlying processes. The appropriate analysis of the data in this kind of research is analysis of variance. The effects in the analysis of variance are attributed to additive processes in performing mental tasks. The present chapter deals with methodological problems inherent in the use of the analysis of variance in RT experiments.

In an analysis of variance, caution is required if the cell frequencies are not equal. The reason for this is that the design in such a case is not orthogonal. In the usual RT experiments, however, the cell frequencies are not equal. Subjects always make errors. Subjects with a relatively large number of errors are usually eliminated, but 5% errors, i.e., 5% missing data, is still very common.

These errors are not randomly distributed among the cells. Usually the number of errors correlates with the length of the RTs for the correct answers: more errors occur in cells with the longer RTs. RT and error rate are generally taken to be convergent measures of item difficulty [1.15]. As in the present experiments the number of errors was small, no special attention was paid to the correlation as such.

Because of this correlation, a weighted means analysis of variance, e.g., a least squares solution, is not appropriate. In a weighted means

---

1 This chapter was written jointly with Dr. L.G.M. Noordman. The authors are indebted to Dr. W. Molenaar for several discussions on topics of this chapter.

analysis, each mean is weighted in proportion to the number of observations on which a mean is based. The means of the cells with the largest number of missing data are underweighted and these are, as has been argued, usually the cells with the longer RTs.

The situation calls for an unweighted means analysis which can more appropriately be termed an equally weighted means analysis. One way to accomplish this is to estimate the missing data. This estimation procedure usually is based on the marginal means. The problem of this procedure is that the estimated scores will be underestimated to the extent that the number of errors correlates with the length of the RTs. A different solution is discussed later on.

Another kind of problem is related to the distribution of the RTs. The analysis of variance assumes that the scores in the cells are derived from normal populations with a constant variance. The RT distribution, however, is not normal but skew with a right tail, and the cell variances might very well be heterogeneous. One may have great confidence in the robustness of the analysis of variance because the analysis is rather insensitive to departures from these assumptions if the cell frequencies are equal [Ref. 2.1, p.331 ff.; Ref. 2.2, p.378 ff.]. But in the usual RT experiments the cell frequencies are not equal, and the question is how robust the analysis is for differences in cell frequencies, even if they would be moderate.

One way to satisfy the assumption with respect to be homogeneity of variances is to transform the scores. It is not uncommon for a transformation not only to make the variances homogeneous but also to normalize the distribution [2.3]. There are, however, some problems inherent in the use of transformations. A logarithmic transformation, which is a rather usual transformation, can serve as an example. A logarithmic transformation is a nonlinear transformation. It differentially affects the scores: higher scores are reduced more than lower scores. Consequently, a logarithmic transformation may introduce or mask interaction effects. The problem associated with the use of transformations is, more fundamentally, that the transformed scale differs from the untransformed scale. This topic has also, but in this case extensively, been treated in the framework of conjoint measurement theory in which techniques have been developed for determining whether any monotone transformation exists which yields, e.g., additive factors on an new scale [2.4,5]. Conclusions obtained on the analysis of variance on transformed scores cannot be generalized to the scores on the untransformed scale if a nonlinear transformation has been used. If a linear model is valid for the logarithmic-transformed scores then the model for the untransformed scores is multiplicative.

From a statistical point of view this model is correct. The question is, however, whether it is sound psychologically. In the present research one is interested in an additive model on the RTs. One can, of course, compute the parameters for an additive model on the RTs that correspond to the significant parameters of the multiplicative model. However, effects that are significant in the multiplicative model need not be significant in the additive model and vice versa, as has been argued above. Consequently, a logarithmic transformation cannot be used when one is interested in the description of mental processes in real time.

The solution that is adopted in the present research is to compute a measure of central tendency and to apply the analysis of variance to these measures. With respect to the problem of missing data, it is clear that this solution is an unweighted means analysis, as is required. With respect to the normality of the distribution and the homogeneity of the variances it should be noted that indices of central tendency computed over RTs approximate a normal distribution better than do the RTs themselves. Analogous to the above cited observation of KEITH SMITH, one may expect also that the variances are more homogeneous.

The experiments have been performed in such a way that for each treatment or treatment combination an index of central tendency can be computed. Calling a treatment or treatment combination an item type, each item type was presented by several item tokens in all the experiments. In some experiments the item tokens for a particular item type are identical. This is the case if one is interested only in these item tokens. In other experiments the item tokens differ, depending on the language material by means of which they are expressed. This is the case when one wishes to generalize to other tokens as well. In the former kind of experiment, language material is a fixed factor. This is the sort of research that CLARK [2.6] termed the method of single cases. In the latter kind of experiment, language material is treated as a random factor.

As an index of central tendency one could use the mean. The index that is computed in the present research, however, is the median. One reason for this is that the distribution of the RTs is skew. The median is a better characterization of the central tendency than the mean in a skew distribution. The medians are also particularly useful to get rid of the effect of outliers. One does not have to bother about criteria and procedures for eliminating outliers. The distribution of the medians will approximate a normal distribution better than the distribution of means does. And, supposedly, the variances might be more homogeneous when using medians than when using means.

If one wishes to generalize beyond the language sample used, the testing procedure requires the use of F' [Ref. 2.7, p.375] or *min* F' [2.6]. What has to be computed is the variance due to the items within treatments or to the interaction of items with treatments, depending on the design. Except for the use of medians instead of means, the procedure to compute this variance is analogous to the one described by CLARK [2.6]. The variances are weighted according to the number of subjects on which an item token score is computed and according to the number of item tokens on which a subject score is computed. In the present study the statistic *min* F' is computed, rather than F', if generalization beyond the language sample is required. The reason is that the original RT data enter in the computation of F', which is undesirable as was discussed above. Not only the values of *min* F' are reported but also the values of F_1. The reason is that *min* F', just as well as F', is a conservative test in particular cases [2.8]. Moreover, the procedure to treat language materials as a random factor is not above dispute [2.9,10]. The interpretation of the results, however, is based on the results of *min* F'.

Some misconceptions with respect to the use of medians have to be discussed. Medians are not additive; consequently, it would not be justifiable to use them in an additive model. The additive models in the present experiments, however, are not based on medians but on means of medians, and means are additive. But then one might object that only generalizations with respect to medians of RTs can be made and not with respect to the original RTs. The answer is that because of the skewness of the distribution it is questionable as to whether one wants to generalize to the original scores. Anyhow, when using means, one would not be able to generalize to the original RTs because of the same skewness, either. This is true whether one considers the RT as a sum of several constant parameters and one error component or as a sum of several stochastic variables.

Some final remarks should be made concerning the assumptions with respect to the variance-covariance matrices in repeated measurements designs. The assumptions with respect to the variance-covariance matrices [Ref. 2.7, p.522; Ref. 2.11, p. 256 ff.] are important for testing overall hypotheses of zero treatment effects that are not based on advance knowledge. A priori comparisons can be tested with the conventional analysis even if the variance-covariance requirements are not met. The structure of the studies to be reported is such that later experiments are usually based on results of earlier experiments. So, predictions in later experiments usually are derived as a priori predictions from earlier experiments, rather than being hypotheses made without advance knowledge of the effects. Consequently, the structure of the variance-covariance matrices is not critical for these predictions. Nevertheless, all the statistical results in the present study are evaluated by means of a procedure that takes the possible violations of the variance-covariance requirements into account. If the variance-covariance matrices do not meet the assumptions, the hypotheses can be tested by means of the Geisser-Greenhouse procedure [Ref. 2.7, p.523 ff.]. According to this procedure, the degrees of freedom for the numerator and denominator are divided by a factor that varies between one and the number of levels in the factor to be tested minus one. The size of this factor depends on the extent to which the variance-covariance matrices do not meet the requirements. Setting this factor to its lower bound, by which one avoids the rather laborious computation of the exact value, tends to make the testing procedure very conservative. The usual tests give results closer to the nominal significance levels than do the results under this conservative Geisser-Greenhouse procedure. Nevertheless, all the significant results reported in the present study remain significant under the Geisser-Greenhouse procedure with the lower bound for the degrees of freedom.

Part I

A Model for Verifying Semantic Relations

Chapter 3 The Selection of the Material from the Hierarchical Domain

In the experiments reported in Chap. 4, subjects had to judge whether two word are same or different with respect to a prespecified category, to be called the target category. Different experimental conditions are defined by the different combinations of the words, as will be explained in this chapter. The material in the experiments consisted of categories and exemplars. The present chapter deals with the selection of the categories and the exemplars.

Some general objectives guided the selection of the material. It was decided to select a rather limited sample of categories and exemplars. In this way all the words could be used in all experimental conditions. The experimental conditions were matched with respect to the words themselves. Experimental results, therefore, cannot be attributed to idiosyncracies of the different words used in different conditions. A second objective was to select common and well-known words that occur frequently in language. These will be found especially among short words. The reason for selecting common words was that one is not interested in peculiarities of the words. Finally, the present experiments deal with hierarchically organized material. The object of study is the subjective hierarchy, e.e., the organization in semantic memory. It was decided to select words that are hierarchically organized in common usage, but also in more technical usage. This was to decrease the possibility that the words have a different organization for different subjects.

Besides these restrictions, the categories to be selected and their exemplars had to satisfy several other criteria. The categories themselves had to be well known just as their names, as was pointed out above. Mammal, for example, is not a common name ("did you see a mammal in the garden?") and, therefore, not an appropriate category. Moreover, the categories to be selected had to be categories of many high-frequency exemplars. Finally, the categories had to be two by two subcategories of two higher order categories so that the material had a hierarchical structure with three levels. This hierarchy yields the possibility of constructing different kinds of relations between concepts. The following four categories were selected: *tree, flower, bird,* and *insect.*

Because of the objective of using common well-known words, a frequency measure should be taken into account in the selection of the exemplars. Consultation of the frequency counts of de la Court in [3.1] and of UIT DEN BOOGAART [3.2] was less informative than required. For example, the latter does not even include in the list of lemmata such supposedly common words as *oak, tulip,* and *sparrow*. Furthermore, category norms, like those of BATTIG and MONTAGUE [3.3] and categorization norms, like those of LOFTUS and SCHEFF [3.4], are not available for the Dutch language. Therefore, an experiment was conducted in which subjects were required to produce exemplars belonging to prespecified categories. The exemplars and especially their frequencies of occurrence provided useful information for the selection of appropriate exemplars. It should be noted that this frequency of occurrence can be conceived of as an indication of the strength of the relation between the category and the exemplar. This relatedness is not necessarily equal to the inverse relatedness between the exemplar and the category [3.4]. This inverse relation was considered in the other three experiments in this chapter.

3.1 Experiment 1: The Production of Exemplars

Exemplars were obtained for the four experimental categories *tree, flower, bird,* and *insect* as well as for 30 other categories. In order to minimize sequential and contextual effects, 10 different random lists of the 34 categories were constructed with the following restrictions. The occurences of each experimental category in the 10 lists were distributed equally among all parts of the lists; the sums of the rank orders for the positions of these categories in the lists were approximately equal. The order of the experimental categories with respect to each other was different for the different lists. Two experimental categories were always separated from each other by at least two other categories. The 34 categories were presented one after an other by means of an audiotape. The presentation of each category was followed by a 20 s time interval, in which the subject had to produce as many members of the category as he could, whichever came into his mind. Two seconds before each new category name, a warning signal was presented in order to notify the subject that he had to stop producing exemplars of the previous category, and that he had to direct his attention towards the category that would follow. The responses of the subject were spoken, and recorded on tape. The subjects were recruited from the same groups from which the subjects were selected for the experiments in the following chapters. The subjects were 100 students, the majority of them undergraduates in psychology. They volunteered in the experiment. Each of the ten lists of categories was presented to five male and five female subjects.

Results. Only the results of the four experimental categories are presented. The norms for all categories were reported in [3.5]. The total number of responses that were given to the four categories was 2548; accordingly, the subjects produced on the average 6.37 exemplars for each category.

Table 3.1. Production frequency indices for exemplars of the categories *tree, flower, bird,* and *insect* (Experiment 1)

	Total frequency	Mean frequency	Standard deviation	Number of different exemplars
(a) *Total set*				
tree	631	12.1	19.47	52
flower	574	6.1	10.91	94
bird	804	7.7	12.22	104
insect	539	12.5	18.83	43
(b) *Production frequency at least 10*				
tree	572	35.8	21.24	16
flower	440	22.0	15.39	20
bird	602	24.1	15.18	25
insect	499	29.4	21.19	17
(c) *Selected exemplars*				
tree	282	47.0	28.40	6
flower	207	34.5	23.19	6
bird	209	34.8	26.52	6
insect	196	32.7	25.57	6

The production frequency data for the four categories are presented in Table 3.1. The number of different exemplars for each category, combined over all subjects, varies from 43 (*insects*) to 104 (*birds*) (Table 3.1, panel a). The mean frequency for each exemplar varies from 6.1 (*flowers*) to 12.5 (*insects*). Many of these words were mentioned by only one subject. Some of them are not members of the category but incidental other associations to the category. If only those exemplars are considered which were produced by at least 10 subjects, the number of different exemplars is reduced in range, from 16 (*trees*) to 25 (*birds*) (Table 3.1, panel b). The mean frequency for the exemplars with at least a frequency of 10 ranges from 22.0 to 35.8.

All exemplars that were used in the experiments of Chaps. 4 and 5 come from the group of exemplars that have a production frequency of at least 10. Six exemplars were selected from each category. In the selection of these exemplars the number of letters and the number of syllables of the words were taken into account. Although short words are to be preferred because of their high frequency, some words were chosen with two syllables from these categories, because there were only two monosyllabic flower names in the group with production frequency above 10. In this way a reasonable cross-category comparability of syllable length was obtained. Furthermore,

Table 3.2. The experimental material: exemplars and categories

boom *tree*	bloem *flower*	vogel *bird*	insect *insect*
eik *oak*	tulp *tulip*	mus *sparrow*	vlieg *fly*
beuk *beech*	roos *rose*	meeuw *gull*	mier *ant*
berk *birch*	narcis *daffodil*	kraai *crow*	wesp *wasp*
spar *spruce*	anjer *carnation*	zwaluw *swallow*	vlinder *butterfly*
wilg *willow*	lelie *lily*	duif *pigeon*	luis (spin)[a] *louse (spider)*[a]
linde *linden*	iris *iris*	valk *falcon*	rups *caterpillar*

[a] Although the *spider* is not an *insect*, it was one of the responses with the highest production frequency to the stimulus *insect*. However, when the Institute for Experimental Psychology, University of Groningen, moved to the Biological Center, it seemed advisable to replace this exemplar of the subjective lexicon by an exemplar of the "objective" lexicon. That is how the *spider* became a *louse*.

the intention was not to select any words that are ambiguous: the word *els* is a tree (*alder*) as well as a girl's name; the word *es* is a tree (*ash*) as well as a musical note and the name of a letter; the word *bij* is an insect (*bee*) as well as a preposition. Resemblance with respect to pronunciation and orthography was also taken into account: not both *mus* and *mees, anjer* and *aster*. The set of words that finally was selected, however, did not entirely meet the last two requirements. Furthermore, only words with a regular spelling were selected: *dahlia* and *crocus* do not meet this requirement. No words were selected that are qualifications of the category, e.g., *appelboom* (*apple tree*).

The exemplars that were finally selected are presented in Table 3.2. The production frequency data (Table 3.1, panel c) show that the mean frequency for each of the exemplars in the categories ranges from 32.7 to 47.0.

3.2 Experiment 2: The Grouping of the Exemplars

The frequency measures of Experiment 1 indicate the strength of the relationship between the category and the exemplars of that category. The present experiment deals with the same relation, approached, however, from the side of the exemplars. The question is whether or not the selected exemplars will be grouped into four main groups, corresponding to the categories that elicited the exemplars in the previous experiment. Similarly, does the grouping of the words give evidence for the two higher ordered categories, corresponding to *plant* and *animal?*

These questions were studied for the selected exemplars of the categories *tree, flower, bird*, and *insect* by means of a sorting experiment, as described by MILLER [1.8,3.6]. The words presented in Table 3.2 constituted the material. There were actually 25 words: both *spider* and *louse* were used, and *maple* instead of *linden* was used. Each word was typed in the center of a 6 x 4 cm card. The 25 cards were presented to the subjects. The subjects were instructed to sort those words together that are similar in meaning according to the subject's overall idea. The subjects were free to select the number of groups as well as the number of words in each group down to a minimum of one. Before a subject started sorting the cards, he had to read all the cards. Fifty subjects, mainly undergraduates in psychology, volunteered in the experiment.

Results. The data were analyzed with JOHNSON's hierarchical clustering techniques [3.7]. The results are presented in Fig. 3.1. The diameter method

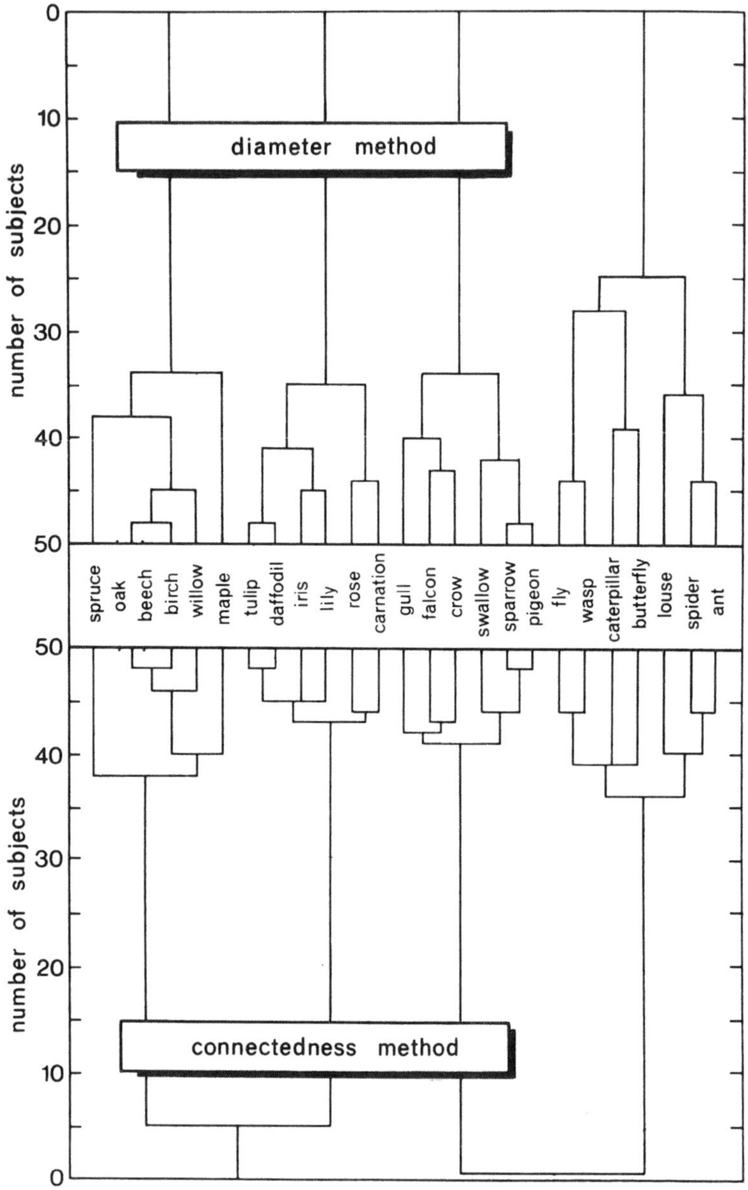

Fig. 3.1. Tree graphs of cluster-analysis for exemplars of the categories *tree*, *flower*, *bird*, and *insect* (Experiment 2)

and the connectedness method yield the same structure almost perfectly. Subjects cluster the words into the four groups of *trees, flowers, birds,* and *insects*. The clustering is very clear: 99.1% of all word pairs that were judged as similar consisted of words of the same category. Because of the low percentage of violations of the category boundaries, almost no evidence could be obtained for the two superordinate clusters.

3.3 Experiment 3: The Categorization of the Exemplars

The present experiment deals, just as the previous one, with the relation of the selected exemplars with the categories. Subjects had to produce the categories of the exemplars. The experiment is the reverse of Experiment 1.

The selected exemplars from the categories *tree, flower, bird,* and *insect* (Table 3.2) were used. The exemplars were presented one at a time by means of a slide projector. Subjects were instructed to produce a category of the presented exemplar. They had to give the category that came first to their minds. They had to give the answer as fast as possible; the response had to be one single word. Two exemplars from the same experimental category (*tree, flower, bird, insect*) were not presented consecutively. The order of the presentation of the exemplars was different for each subject. Each exemplar occurred at a different place in each list; the sums of the rank orders for the position of the exemplars in the lists were about equal for all exemplars. Within these restrictions, the order of presentation was randomized.

Twenty students, mainly undergraduates in psychology, served as the subjects. They volunteered in the experiment.

Before presenting the results, one restriction with respect to the interpretation of the results should be kept in mind. Because of the fact that exemplars of only four categories were presented, contextual effects might have influenced the results considerably. The way in which a previous exemplar was categorized will certainly influence the way in which a subsequent, similar exemplar will be categorized.

Results. The categories that were produced were predominantly the categories that had served as stimuli in the exemplar-production experiment (Experiment 1): 87% of the answers were the category names *tree, flower, bird,* and *insect* (Table 3.3). Nine percent of the answers were supercategory names: categories of the next higher level in the hierarchy. These 9% were almost exclusively the supercategory *animal* as response to exemplars of *insects*. As is indicated in Table 3.3, 34% of the answers to *insect* exemplars were the supercategory name and 61% were the category name. Two percent of the answers were subcategories: categories at a lower level than *tree, flower, bird,* and *insect*, e.g., *cut flower* and *bird of prey* (which in Dutch are expressed in one word). Finally 1.5% of the responses were false or just silence.

Table 3.3. Categorization responses [in percentages for exemplars of *tree, flower, bird,* and *insect* (Experiment 3)]

Exemplars of	Responses			
	Sub-category	Category	Super-category	False/no
tree	1	99		
flower	4	93	2	1
bird	3	96	1	
insect		61	34	5

The conclusion is that the category names *tree, flower, bird,* and *insect* are the "natural" categories for the exemplars that were selected for the experiments to be reported. For the *insect* exemplars, too, the category *insect* is the answer with the highest frequency.

Because almost no evidence was obtained thus far on the three-level organization of the material, an experiment was performed in which pairs of words were presented. Before reporting this experiment, the types of relations between two words in a pair that are derived from the hierarchical organization of the material will be discussed.

3.4 Semantic Relations Between Two Exemplars

The task in the experiments, reported in Chap. 4, is to judge whether two words are same or different with respect to prespecified target categories. Several types of relations with respect to the hierarchical organization of the material can be distinguished between two words in a pair. These relations are illustrated in Table 3.4 for the exemplars of the categories *tree, flower, bird,* and *insect* with respect to the target category TREE.[1] The relations are defined with respect to the categories at the same level as the target categories in the experiment. A word pair is called a same word pair if both words belong to the same category on the level of the target category. A word pair is called a different word pair if both words do not belong to the same category at the level of the target category.

Table 3.4. Relations between the words

Example of item[a]		Targets	TREE FLOWER BIRD INSECT	PLANT ANIMAL
oak	beech		same	same-related
oak	rose		different-related	same-unrelated
oak oak	sparrow ant		different-unrelated	different

[a] The examples illustrate the relations with respect to the target categories TREE and PLANT, respectively.

It depends on the hierarchical status of the target category whether one can distinguish between a same-related and a same-unrelated word pair. The distinction between same-related and same-unrelated word pairs is made if the target categories are in fact supercategories, viz., PLANT and ANIMAL. A word pair is called a same-related pair if both words belong to the same supercategory, and also to the same category. If the two words belong to the same supercategory but not to the same category, the pair is called a same-unrelated pair.

A distinction between different-related and different-unrelated word pairs can be made if the target categories are at the same level in the hierarchy as the categories (e.g., TREE). The distinction between differ-

[1] When a category name refers to the target category it is capitalized in this part.

ent-related and different-unrelated word pairs can be described in terms of a higher order category that is common for words in different-related pairs but not common for words in different-unrelated pairs. For the material discussed in this chapter this comes down to the following: a word pair is different-related if both words do not belong to the same category but do belong to the same supercategory; a word pair is different-unrelated if the words belong neither to the same category nor to the same supercategory. These types of word pairs will be referred to as item types.

3.5 Experiment 4: The Categorization of Pairs of Exemplars

Experiments 2 and 3 did not present evidence for the organization of the material in three hierarchically related levels. The reason appears to be the saliency of the categories *tree, flower, bird,* and *insect*: these are the dominant responses in Experiment 3. Moreover, the number of violations of the category boundaries in Experiment 2 was only very small. Evidence for higher order categories can perhaps be obtained from a categorization task in which the categories *tree, flower, bird,* and *insect* are not or at least not always the appropriate response.

The items in the present experiment were pairs of words, the words being arranged horizontally. The items were presented by means of a slide projector. Subjects had to produce the name of a category common to both words. They had to give the answer as fast as possible and the response had to be a single word.

The words that were used were the selected exemplars from the categories *tree, flower, bird,* and *insect* (Table 3.2). Because of the saliency of the level of the categories (*tree, flower, bird,* and *insect*), word pairs were categorized with respect to these categories in same, different-related, and different-unrelated item types. Six same word pairs were selected from each category, yielding 24 same items. Twelve different-related word pairs were selected from each of the supercategories *plant* and *animal,* yielding 24 different-related items. Finally, there were 12 different-unrelated word pairs. Each word occurred approximately an equal number of times in each item type and each word occurred approximately an equal number of times in the left-hand position and in the right-hand position in a pair. The material was presented in five blocks of 12 items each. The blocks were balanced with respect to all the variables just mentioned. The sequence of items in a block was such that no word occurred more than once in three consecutive items. There were two different orders of the items within the blocks, and five different orders between the blocks, resulting in ten different orders. The 20 students of Experiment 3 served as the subjects.

Table 3.5. Categorization responses (in percentages) for pairs of exemplars of *tree, flower, bird,* and *insect* (Experiment 4)

Item type	Word pair of exemplars of	Responses: Category	Super-category (flying)	Super-category (plant/animal)	Higher order category (living/nature)	False/no
same	tree and tree	98		1		1
	flower and flower	85		13	2	
	bird and bird	95		4	1	
	insect and insect	56	1	40	2	2
different-related	tree and flower			51	46	3
	bird and insect		10	55	33	2
different-unrelated	tree or flower and bird or insect				94	6

Results. The results are presented in Table 3.5. The responses for the same items were predominantly the categories that had served as the stimuli in the exemplar production experiment (Experiment 1): 84% of the answers were the category names *tree, flower, bird,* and *insect*. This result is in agreement with the results of the categorization experiment of the exemplars (Experiment 3).

The more interesting items are the different-related items. The majority of the answers (53%) were the supercategories *plant* and *animal*. The supercategory *flying* (*animals*) was given as the response for another 5% of the items. The higher order categories *living* (*thing*) or *nature* were produced as the response for 40% of the different-related items. These higher order categories are the lowest common categories for words in the different-unrelated word pairs. Accordingly, 94% of the responses for the different-unrelated items were these higher order categories.

The conclusion of the present experiment is that the supercategories *plant* and *animal* are natural supercategories of the selected exemplars, as is indicated by the results of the different-related items. Although a considerable percentage of responses was constituted by the higher order categories, the present experiment provided evidence for the three-level hierarchical organization of the material. One may conclude from these four experiments that the selection of the exemplars of the categories *tree, flower, bird,* and *insect* satisfied the criteria that were adopted.

Exemplars from two other categories - *time* and *weight* - were used only later on in the series of experiments. Production frequency data are presented at the appropriate place (Sect. 4.3).

Chapter 4 The Verification of Hierarchical Relations

This chapter aims at developing a model that can account for the results in the semantic information processing experiments which were discussed in the Introduction. The first task in the present study is to check whether these findings can be replicated reliably. These findings can be summarized as follows. First, the RT for same judgments is shorter the more related the words are; this will be called the semantic distance effect for same judgments. Second, the RT for different judgments is longer the more related the words are; this will be called the semantic distance effect for different judgments. Finally, the data presented in many experiments suggest that the RT for same judgments is shorter than for different judgments; this will be called the same/different effect. The last two effects were tested in Experiment 1. A model of the retrieval processes in a hierarchical network is presented that can also account for the first effect. The first effect was tested in Experiment 2.

Predictions based on this model were further tested in an experiment in which the difference between the concepts was increased (Experiment 3) and in an experiment with a different number of target categories (Experiment 4).

4.1 Experiment 1

Pairs of words were presented, the words being arranged horizontally. The subject had to decide whether both words were same or different with respect to one of the four target categories TREE, FLOWER, BIRD, and INSECT. All words were exemplars of these categories. The relation between words in a pair can be classified as same, different-related, or different-unrelated (Table 3.4). Besides word pairs in which both words were exemplars, word pairs were used in which one word was an exemplar and the other word one of the four prespecified categories. The former type of pairs is identified as Ex-Ex pairs, the latter as Cat-Ex pairs. The Cat-Ex pairs can be subdivided into same, different-related, and different-unrelated word pairs, just as the Ex-Ex pairs. Examples of the items are presented in Table 4.1.

Table 4.1. Examples of item types and their frequency of occurrence in Experiment 1

	Same	Different-related	Different-unrelated
Ex - Ex	oak beech (24)	oak tulip (12)	oak ant (12)
Cat - Ex	oak tree (20)	oak flower (10)	oak bird (10)

Four predictions were tested. First, the RT for same judgments is shorter than for different judgments. Second, the RT for different judgments is longer the more related the words are. It was assumed that the comparison process is a process of retrieving the categories of the words to be compared and that the first step in comparing two exemplars is the retrieval of the categories of the words. The processes underlying the Cat-Ex comparisons must, therefore, be very similar to those underlying the Ex-Ex comparisons. Consequently, the third prediction was that the semantic distance effect for different judgments shows up in Cat-Ex pairs as well as in Ex-Ex pairs. This prediction is opposed to the claim of SCHAEFFER and WALLACE [1.27], as was discussed in the Introduction. Finally, for the same reason, the RT for Cat-Ex pairs was predicted to be shorter than the RT for Ex-Ex pairs.

4.1.1 Method

Material. The experimental material consisted of 88 pairs of words: 48 Ex-Ex word pairs and 40 Cat-Ex word pairs. The design was balanced with respect to the number of same and different items as well as to the number of different-related and different-unrelated items. The number of items for each of the six item types is indicated in Table 4.1. Each item type can be constructed in several ways depending on the material; there are, for instance, four possible Ex-Ex same items, as the two exemplars can both be a *tree*, a *flower*, a *bird*, or an *insect*. Each item type was balanced with respect to these possible realizations in the material. Also, each item type was balanced with respect to the number of exemplars from each category. Each word occurred approximately an equal number of times in each item type and each word occurred an equal number of times in the left-hand position and in the right-hand position in a pair. This was approximately the case for each of the six item types. The material was presented in blocks of 16 items each. Each block was balanced with respect to all the variables just mentioned. The sequence of items was such that not more than three consecutive items required the same answer and that no word occurred more than once in three successive items. Within these restrictions the order of presentation of the items was randomized. The three experimental blocks of the Ex-Ex items were preceded by one practice block. There were two and one-half experimental blocks of Cat-Ex items preceded by 16 practice items. The practice blocks were in all respects similar to the experimental blocks. All the words were presented at least once in a practice block. The frequency of occurrence of the exemplars in a practice block was the same for the four categories.

Procedure. The items were presented using a Leitz slide projector. The distance of the subject to the screen was approximately 2.5 m. The items subtended a visual angle of about 14 deg in a horizontal direction. The subject had to press one of two response keys, one for the response "same" and one for the response "different". The time was recorded in ms from the onset of the stimulus to the moment the subject pressed a key. During the experiment the subject kept his index fingers on the keys. The subject was seated in a soundproof room, and the projector and timer were located in an adjacent room. The slides were projected through a window of double, plan parallel coated glass in the wall. The subject was told that on each slide he would see two words, each word being the name of a *tree, flower, bird,* or *insect* or the word *tree, flower, bird,* or *insect* itself. The question he had to answer was whether both words refer to one of the four prespecified categories: TREE, FLOWER, BIRD, and INSECT. If both words referred to the same category he had to press the right key, if they referred to difrerent categories he had to press the left key. The same answer-key assignment was used throughout, because in an earlier experiment (unpublished data) in which one-half of the subjects had to press the right-hand key for "same" and the other one-half of the subjects the left-hand key for "same", the assignment of "same" and "different" responses to the keys did not interact with any of the other factors.

The subject was required to give the correct answer as quickly as possible. If, however, he realized that he had made an error, he had to say "wrong", and the experimenter delayed the presentation of the next item by about 2 s. At the beginning and at the end of each block a warning signal was given. The time between two blocks was approximately 1 min. The time between two items was 2 s.

In order to make the subject aware of the hierarchical organization of the material, the experimenter mentioned the words one after the other in the instruction and the subject was required to say to which category each word belonged. After the practice block, the subject was asked whether there were any problems, and the subject had to give the names of the four target categories. To one-half of the subjects the Ex-Ex blocks were presented before the Cat-Ex blocks; for the other half the order was reversed. The experiment lasted about 25 min.

Subjects. Twenty-six students, the majority of them undergraduates in psychology, volunteered in the experiment.

4.1.2 Results

Error rates were on the average 3.9%. Only the RTs for correct responses were used in the analysis of the data. For each subject, six medians were computed, one for each item type: same, different-related, different-unrelated for both Ex-Ex and Cat-Ex pairs. The means of these medians are presented in Table 4.2. These medians were analyzed by means of an analysis of variance with one between subjects factor, viz., the order of presentation of the Ex-Ex and Cat-Ex items. There were two within subjects factors: Ex-Ex items vs Cat-Ex items, and same, different-related, and different-unrelated items. Both the F_1 and the *min* F' for each effect are reported. The language material that was considered as the statistically random factor consisted of the item tokens within each category or category combination. The variance due to the position of the items in the experiment was, of course, not considered as variance due to the language material as such and was eliminated from this variance.

Table 4.2. Mean RT in ms for the item types in Experiment 1 (target categories TREE, FLOWER, BIRD, INSECT)

	Same	Different-related	Different-unrelated
Ex - Ex	1123	1409	1212
Cat - Ex	1068	1260	1157

The predictions were confirmed by the data. The RT for same judgments is shorter than for different judgments: $F_1(1,48) = 61.42$, $p < 0.001$; $min\ F'$ $(1,117) = 30.84$, $p < 0.001$. There is a semantic distance effect: the RT is longer for the different-related items than for the different-unrelated items: $F_1(1,48) = 39.14$, $p < 0.001$; $min\ F'$ $(1,104) = 18.55$, $p < 0.001$. The semantic distance effect also occurs for the Cat-Ex items separately: $F_1(1,48) = 11.30$, $p < 0.01$; $min\ F'$ $(1,127) = 4.82$, $p < 0.05$. Finally, the RT for Ex-Ex items is longer than for Cat-Ex items: $F_1(1,24) = 4.69$, $p < 0.05$; $min\ F'$ $(1,37) = 3.70$, $p < 0.05$ one tail.

4.1.3 Discussion

The results suggest some principles for a model of the comparison processes. The same/different effect suggests that subjects first try to find positive evidence with respect to the criterion in the task. The fact that a semantic distance effect occurs also for the Cat-Ex items and the fact that the RT for Cat-Ex items is shorter than the RT for EX-Ex items are in agreement with the assumption that the comparison process is a process of retrieving the categories of the words to be compared.

The model that has to account for the data is presented in Fig. 4.1. It is a two-stage model. The first stage is a verification stage: positive evidence with respect to the criterion is searched for. If no positive evidence is found, the process enters a second stage. That stage could be called a falsification stage. It consists of a search for differences between the concepts, in order to justify the conclusion "different". It can be conceived of as a double check.

In the first stage, categories of the words are retrieved and the category of one of the words is tagged as the target in the item (T_i). Which category will be tagged may be the outcome of a racing process: which category is selected first? If the word was already a category, the category is placed directly in a comparator as the target T_i. The category of the other word is placed in the comparator and compared to the target T_i. If a match occurs, the output is "same". The process in the first stage is similar to the model of COLLINS and QUILLIAN for true sentences [1.13]. If no match is found, the process enters a second stage. The aim of that stage is to justify the judgment that the two words do not belong to the same category. Therefore, differences have to be found between the two words. If one has to differentiate between word meanings in a hierarchical domain, the most efficient strategy is a hierarchical decision strategy, starting from top to bottom. Consequently, when a mismatch is found in stage 1, the first step in stage 2 is the retrieval of supercategories. The supercategory of the target category (T_i') is placed in the comparator, as is the supercategory of the other category. If the comparison reveals that the supercategory of the other category is other than T_i', then positive evidence has been found and the answer "different" can be given. The model up to this point describes the process of two different-unrelated words, e.g., *oak sparrow*. However, if the other supercategory matches the supercategory of the target (T_i'), as is the case in the different-related item type (e.g., *oak tulip*), then the difference has to be found in a lower level category: again at the level of the target category. Logically, only a mismatch can occur at this point in the model and the answer "different" has to be given.

When processing different-related word pairs, the categories are searched for again in the falsification stage. The objective of the search, however, differs from the one in the verification stage. Now the process aims at finding differences. Presumably, the information from the search in the verification stage has not been retained. This return to the same operation is analogous to the rechecking of the same constituents in the CARPENTER and JUST [4.1] comparison model.

The two-stage analysis accounts for the difference in RT between the "same" and "different" answers. The properties of the second stage account for the semantic distance effect, i.e., the RT for different pairs is longer the more related the words are. These results cannot be accounted for by some other procedures such as changing a truth index and time-limited search, reported in language processing studies. The changing of a truth index to *false* if no match is found as well as the execution of the answer "different" if within a certain time period no evidence for "same" has been found cannot be reconciled with the finding of the difference in RT between different-unrelated and different-related item types.

When one of the words is a category, the retrieval step for this word of stage 1 can be skipped. Consequently, the RT for Cat-Ex pairs should be shorter than for Ex-Ex pairs, and this was the case. However, the difference in RT between the Cat-Ex and Ex-Ex pairs might be attributed to the higher frequency of the category names as compared to the exemplar names. A decrease in RT due to high frequency of occurrence can be accounted for by spreading of activation [1.20,34]. However, it is not clear how the higher frequency of the category names can account for the following observation which will be discussed.

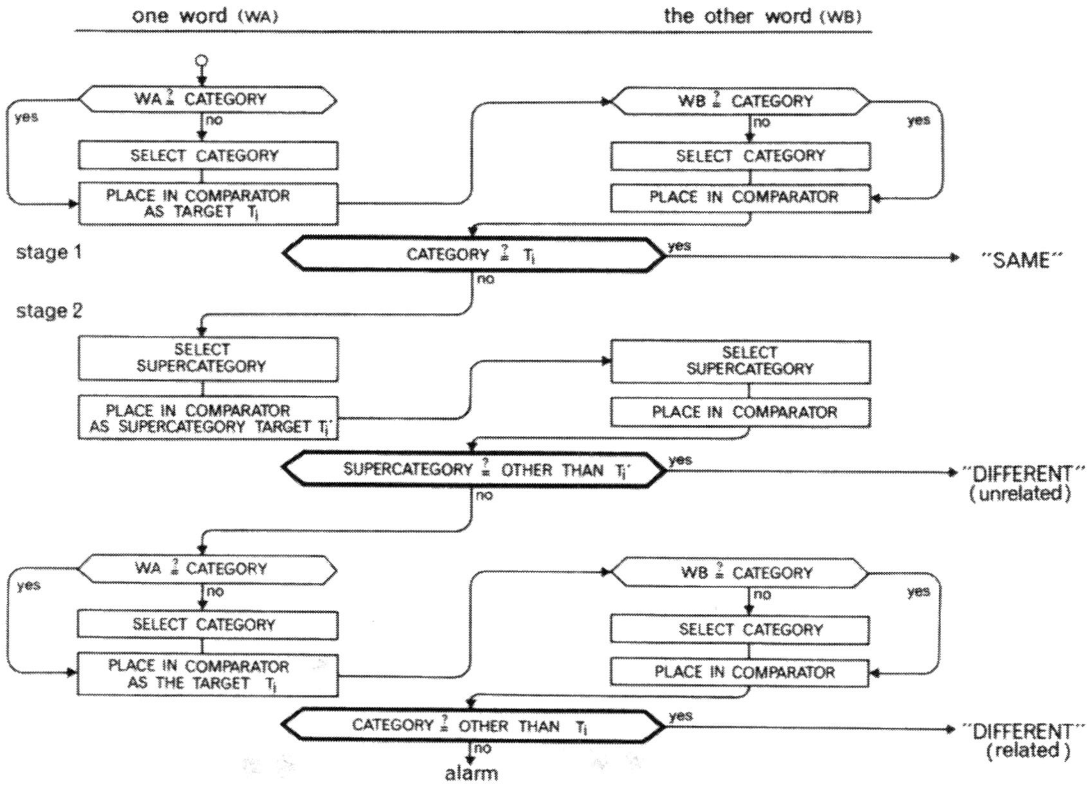

Fig. 4.1. Process model for the verification of hierarchical relations in Experiment 1

When one of the words is a category, a retrieval step can be skipped in stage 1, and also in stage 2 for the different-related items. Consequently, the difference between Ex-Ex and Cat-Ex judgments for same pairs as well as for different-unrelated pairs is one retrieval step, but that difference is two retrieval steps for different-related pairs. The data show that the former difference is indeed smaller than the latter: $F_1(1,48) = 7.28$, $p < 0.01$; *min* F' $(1,101) = 2.89$, $p < 0.05$ one tail. So, the model does also account for the difference in semantic distance effect for Ex-Ex pairs and Cat-Ex pairs.

4.2 Experiment 2

The aim of the present experiment was to test the semantic distance effect for same items. In this experiment the relation between words in a pair could be classified as same-related, same-unrelated, and different (Table 3.4). It was predicted that the RT for same-related items is shorter than the RT for same-unrelated items.

The experiment is similar to the previous one, except that the word pairs had to be judged with respect to two targets: the supercategories PLANT and ANIMAL. The model can easily account for the semantic distance effect for same items by simply adding an extra step to stage 1. Categories are retrieved in the order of the hierarchy: from lower level categories to higher level categories. The RT for a same-related word pair (e.g., *oak beech*) should then be shorter than the RT for a same-unrelated word pair (e.g., *oak tulip*) because in the former case the words are same already with respect to a category of a lower level than the target category. In the latter case, the similarity is not detected until the higher order target category has been retrieved.

Two other predictions are the same as in Experiment 1. First, the RT for same items should be shorter than the RT for different items. Second, the RT for EX-EX items should be longer than the RT for Cat-Ex items.

Finally, from the model one can predict that there will be no difference between the Ex-Ex and Cat-Ex items in the semantic distance effect. The reason is that in comparing Cat-Ex words one and only one step in stage 1 can be skipped for both the same-related and the same-unrelated items.

4.2.1 Method

Material. The items were constructed in a way similar to Experiment 1. The same words were used. The experimental material consisted of 128 items.

The numbers of same-related, same-unrelated, and different items were 16, 16, 32, respectively, for both Ex-Ex items and Cat-Ex items. Each item type can be constructed in several ways depending on the material; e.g., there are four possible different Ex-Ex items: the combination of *tree* or *flower* with *bird* or *insect*. Again each item type was balanced with respect to these possible realizations in the material. Each item type was balanced with respect to the number of exemplars from each category as well.

The Ex-Ex items as well as the Cat-Ex items were presented in four blocks of 16 items each, preceded by a practice block. All blocks were similar to each other in all respects. They were constructed in the same way as the total set of items was constructed. All other restrictions on the material were the same as in Experiment 1.

Procedure. The subjects had to judge whether both words in a pair referred to one of the two prespecified categories: PLANT or ANIMAL. For one-half of the subjects, the Ex-Ex items were presented before the Cat-Ex items. For the other half, the order was reversed. The procedure was the same as in Experiment 1 in all other respects as well.

Subjects. Twenty-six students, the majority of them undergraduates in psychology, volunteered in the experiment.

4.2.2 Results

Error rates were on the average 5.6%. For each subject six medians for the correct responses were computed, one for each item type. The means of these medians are presented in Table 4.3. A three-way analysis of variance was performed on these medians. The between subjects factor was the order of presentation. The two within subjects factors were Ex-Ex vs Cat-Ex items and same-related, same-unrelated, and different items.

Table 4.3. Mean RT in ms for the item types in Experiment 2 (target categories PLANT, ANIMAL)

	Same-related	Same-unrelated	Different
Ex - Ex	1107	1227	1305
Cat - Ex	1061	1168	1286

The predictions were confirmed by the data. The RT for the same-related items is shorter than for the same-unrelated items: $F_1(1,48) = 35.24$, $p < 0.001$; $min\ F'\ (1,144) = 11.72$, $p < 0.001$. The data indicate a semantic distance effect for the Cat-Ex items as well as for the Ex-Ex items. In fact, the semantic distance effect for the Ex-Ex items does not differ from the semantic distance effect for the Cat-Ex items, confirming the prediction: $F_1 < 1$. Furthermore, the RT for the same items is shorter than for the different items: $F_1(1,48) = 87.83$, $p < 0.001$; $min\ F'(1,144) = 29.27$, $p < 0.001$.

One prediction was not quite confirmed by the data. The RT for Ex-Ex items was longer than the RT for Cat-Ex items, but not significantly so, when testing with *min F'*: $F_1(1,120) = 5.98$, $p < 0.05$; $min\ F'\ (1,190) = 2.22$, $p < 0.10$ one tail.

4.2.3 Discussion

In agreement with the predictions, the RT increases from same-related to same-unrelated to different item types. This means that one more step has to be added to stage 1 of the model (Fig. 4.1). If the words do not belong to the same category, both categories have to be processed further. If on the other hand the words belong to the same category, only one category has to be processed further.

The RT increases with the number of operations as indicated by the global articulation of the model in three steps. It is worthwhile to compare these results with those in Experiment 1. The correspondence between the RTs in the two experiments is a correspondence in terms of the number of steps in the model and not in terms of the items. The RTs for the same, different-unrelated, and different-related item types in Experiment 1 are almost identical to the RTs for the same-related, same-unrelated, and different item types, respectively, in the present Experiment 2, although different-unrelated items in Experiment 1 are different items in Experiment 2 and different-related items in Experiment 1 are same-unrelated items in Experiment 2. This correspondence between the two experiments is found both for the Ex-Ex and Cat-Ex items.

The difference in RT between the Ex-Ex and Cat-Ex items is not very reliable. One could argue that, in the course of the experiment, the identification of the category of an exemplar is becoming a process that hardly requires extra time. A pilot experiment suggests that the difference between Ex-Ex and Cat-Ex disappears very rapidly. Three subjects were tested with the material of Experiment 1, once on each of nine consecutive days. The Ex-Ex items were presented before the Cat-Ex items. The difference in RT between same, different-related, and different-unrelated continued to be significant: on the ninth day $F_2(2,82) = 5.22$, $p < 0.01$. The difference between the Ex-Ex and Cat-Ex items, however, nearly disappeared as early as the second day: on the first, second, and ninth days $F_2(1,82) = 20.04$, $p < 0.001$; $F_2(1,82) = 2.85$, $p < 0.10$; $F_2(1,82) < 1.00$, respectively.

4.3 Experiment 3

With respect to the difference in RT between the same and different judgments, one can ask whether the results are an artifact of the material. One might argue that the words in the different-unrelated items in Experiment 1 are closely related, after all. The question is whether the RT for same items is shorter than for different items if the different items are composed of words actually far apart in meaning.

Other models, such as sampling models, were not concerned with the difference in RT between same and different items. In comparing two words in a sampling model, elements are sampled and same elements give evidence for the answer "same", different elements for the answer "different" [1.27]. The answer is given as soon as a threshold is reached. The more related the words are, the faster the threshold for same will be reached, and the slower the threshold for different will be reached. If no difference is postulated between the threshold-value same and the threshold-value different, then no difference in RT can be predicted between same word pairs and different-unrelated word pairs. The observed differences in RT between the same pairs and the different-unrelated pairs in the present experiments should then be accounted for by assuming that the different-unrelated words are in fact rather related.

Table 4.4. Production frequency indices for the exemplars of the categories *time* and *weight*

	Total frequency	Mean frequency	Standard deviation	Number of different exemplars
(a) *Total set*				
time	499	6.7	15.02	75
weight	437	11.2	23.87	39
(b) *Production frequency at least 10*				
time	399	39.9	20.67	10
weight	391	48.9	32.31	8
(c) *Selected exemplars*				
time	176	44.0	23.45	4
weight	307	76.8	13.89	4

Table 4.5. The experimental exemplars of Experiment 3

(boom) *tree*	(vogel) *bird*	(tijd) *time*	(gewicht) *weight*
eik *oak*	meeuw *gull*	uur *hour*	kilo *kilogram*
beuk *beech*	kraai *crow*	minuut *minute*	gram *gram*
spar *spruce*	zwaluw *swallow*	jaar *year*	ons *ounce*
wilg *willow*	duif *pigeon*	week *week*	pond *pound*

The aim of this experiment is to test the generality of the same/different effect. The categories that have been used are *tree, bird, time,* and *weight*. *Time* and *weight* are two out of the 34 categories for which exemplars have been obtained in the production of exemplars experiment (Sect.3.1). Production frequency data for the exemplars of the categories *time* and *weight* are presented in Table 4.4. The exemplars used in the present experiment (Table 4.5) were selected from the group of exemplars that have a production frequency of at least 10. Four exemplars were selected from each category, because there were only four exemplars for the category *weight* with production frequency of at least 10 that met the other selection criteria, formulated in Chap. 3. The mean frequencies of the exemplars are, relative to those of the concrete categories (Table 3.1), very high.

Different-unrelated word pairs can be constructed from the categories *tree, bird, time,* and *weight* in which the words actually are far apart in meaning (concrete vs abstract). The prediction of the model is that the RT for same items (e.g., *oak beech*) is shorter than for these different-unrelated items (e.g., *oak hour*).

4.3.1. Method

Material. Only Ex-Ex word pairs were used in this experiment. The exemplars that were used can be found in Table 4.5. There were three item types: same, different-related (both words are concrete nouns, or both are abstract nouns), and different-unrelated word pairs (one concrete and one abstract noun) (Table 4.6). It should be noted that items such as *oak sparrow* are classified in the present experiment as different-related as opposed to the different-unrelated classification in Experiment 1 of this chapter. This is due to the difference in material in both experiments. Different-related and different-unrelated item types in Experiment 1 were differentiated on the level of the supercategories *plant* and *animal*. In the present experiment, however, one has to make use of higher order categories in order to differentiate between the different item types.

Table 4.6. Relations between the words in Experiment 3

Example of item		Targets	TREE BIRD TIME WEIGHT
oak	beech		same
oak	sparrow		different-related
oak	hour		different-unrelated

The experimental material consisted of 48 items: 24 same items, 12 different-related, and 12 different-unrelated items. The items were constructed according to the same principles and restrictions as in the previous experiments.

The items were presented in three experimental blocks, preceded by one practice block. Each block reflected the organization of the total set of items in the experiment.

Procedure. The items were presented in 12 different orders: the three experimental blocks came in six different orders, and two different orders of the items within the blocks were used. Two subjects were assigned to each one of the 12 orders of presentation. The subjects had to answer the question whether both words refer to one of the four prespecified categories: TREE, BIRD, TIME, and WEIGHT. In all other respects the experiment was similar to Experiment 1.

Subjects. The subjects were 24 students, the majority of them undergraduates in psychology. They volunteered in the experiment.

4.3.2 Results and Discussion

The percentage of errors was on the average 3.6%. For each subject, three medians for the correct responses were computed, one for each item type.

An analysis of variance was performed on these medians. The means of the medians are presented in Table 4.7.

Table 4.7. Mean RT in ms for the item types in Experiment 3 (target categories TREE, BIRD, TIME, WEIGHT)

	Same	Different-related	Different-unrelated
Ex-Ex	928	1118	988

The data confirm the model. The RT for same items is shorter than the RT for different-unrelated items: $F_1(1,23) = 9.37$, $p < 0.01$; $min\ F'\ (1,45) = 3.60$, $p < 0.05$ one tail. The conclusion is that subjects first try to find positive evidence with respect to the criterion. If this evidence cannot be found, a falsification process follows, in which evidence is searched for the answer different.

That a semantic distance effect occurs is clear enough from the data. It should be noted that the slope of the means is similar to the slope of corresponding means in Experiment 1, as it should be.

4.4 Experiment 4

In the previous experiments, words from four categories were used. The same four categories or the two supercategories were the target categories in the experiment. Therefore, each word is an exemplar of one of the targets. The aim of the present experiment was to study the processing of words that are not exemplars of a target category. Therefore, the experiment was performed in the following way. Again, exemplars from the four categories *tree, flower, bird,* and *insect* were used. But there is only one target category at a time, with respect to which a word pair has to be judged. This target category is presented just before the presentation of the word pair. The target category could be TREE or FLOWER or BIRD or INSECT.

The task with only one target is in itself interesting. The results should be in agreement with the model. But the items constructed with words that are not exemplars of the target category are of main interest. It can be argued that the fact that the item as such has no relation to the target is used as a shortcut in the process. Indeed, it seems plausible that at an early stage in the process the item is assessed to be irrelevant to the criterion. Consequently, the decision about items in which both words are not related to the target might be made relatively quickly: a "different" answer in this case would be shorter than a "same" answer!

Another problem that should be studied in the task with only one target category is whether the words in an item are processed in a left-to-right order. Once it has been detected that one of the words is not an exemplar of the target category, the other word does not have to be processed anymore. If the words are processed in a left-to-right order, the RT for a word pair in which the word at the left-hand position is an instance of the target category and the word at the right-hand position is not should be longer than the RT for items with the opposite order. Such an effect was found by SHOBEN [1.29] in a word-comparison task in which the words were arranged vertically. It should be noted that the subjects explicitly were instructed to read the top word of the word pair first.

In the previous experiments the relation of the words to the target was determined by the relation between the two words. More specifically, with respect to whichever target category an item finally is judged, the relation between one of the two words and the target category is always the same as the relation between the two words.

However, when words have to be judged with respect to only one prespecified target category, the relation of each word to the target is not determined by the relation between the two words. On the other hand, the relations of the words to the target in almost all cases determine the relation between the words. Consequently, the word pairs were classified according to the relation of the words with respect to the target category and moreover, only in some cases, according to the relation between the words.

The relation of a word to the target category (e.g., TREE[1]) is a same relation if that word belongs to that category, e.g., *oak*. The relation of a word with the target category is a different-related relation if the word does not belong to that category, but if word and category belong to the same supercategory, e.g., *tulip*. The relation is a different-unrelated relation if the word belongs neither to the target category nor to the higher order category of the target, e.g., *sparrow*. Table 4.8 illustrates the item types of the present experiment.

---

1 The examples in the text on Experiment 4 assume that the target category is TREE.

Table 4.8. Item types illustrated with respect to the target TREE and their frequency of occurrence in Experiment 4

Examples of word pairs	Relation to target of Left-hand word	Relation to target of Right-hand word	Relation between words	Frequency
1) oak beech	same	same	same	54
2) oak tulip	same	different-related	different-related	6
3) tulip oak	different-related	same	different-related	6
4) oak sparrow	same	different-unrelated	different-unrelated	6
5) sparrow oak	different-unrelated	same	different-unrelated	6
6) sparrow pigeon	different-unrelated	different-unrelated	same	6
7) sparrow ant	different-unrelated	different-unrelated	different-related	6
8) tulip rose	different-related	different-related	same	6
9) sparrow tulip	different-unrelated	different-related	different-unrelated	6
10) tulip sparrow	different-related	different-unrelated	different-unrelated	6

The following predictions were made. The predictions for the items in which at least one word is an exemplar of the target category are the same as those in the previous experiments: the RT for same items should be shorter than for different items; the RT for different-related items should be longer than for different-unrelated items.

The RT for items in which both words have a different-unrelated relation to the target category (e.g., *sparrow ant*) should be shorter than the RT for items in which one word has a same and the other one a different-related or different-unrelated relation to the target category (e.g., *oak tulip* and *oak ant*, respectively). Finally, the RT for items in which both words have a different-unrelated relation to the target category (e.g., *sparrow ant*) should be shorter than the RT for items in which both words have a different-related relation to the target category (e.g., *tulip rose*) and shorter than the RT for items where one word has a different-related and the other a different-unrelated relation to the target category (e.g., *sparrow tulip*) .

No prediction with respect to the left-to-right order of processing was apparent in advance. If the RT for items in which the first word has a same relation to the target category and the second word does not (e.g., *oak sparrow*) turns out to be longer than for items in which the second word has a same relation to the target category and the first one does not (e.g., *sparrow oak*), then one may assume a left-to-right processing order. If, however, the former RT turns out not to be different from the latter RT, one may conclude that there is no left-to-right processing of the item or at least that, if there is such an order, both words are processed even in case the first word is not an exemplar of the target.

4.4.1 Method

Material. The 10 different item types and their frequency of occurrence in the experiment are presented in Table 4.8. The experimental material consisted of 108 items: 54 same items and 54 different items. Of the 54 same items in the experimental blocks, 42 were fillers. The items were constructed according to the same principles and restrictions as in the previous experiments. The items were presented in six blocks of 18 items each; these blocks were preceded by two practice blocks. The first item in each block was a practice trial.

Procedure. At each trial the target category was presented for 2 s. It was followed by the presentation of the word pair. An item was followed by a blank interval of 4 s. All targets occurred an approximately equal number of times in each block. No particular target was presented in two consecutive trials. The order of presentation of the items was the same for all subjects. In all other respects the experiment was similar to the previous ones.

Subjects. The subjects were 20 students, the majority of them undergraduates in psychology. They volunteered in the experiment. Half of the subjects had to press the right-hand key for the answer "same", the other half had to press the left-hand key for the answer "same". In this experiment, all subjects were right-handed.

4.4.2 Results

The number of errors was on the average 5%. For each item type a median for the correct responses was computed for each subject. An analysis of variance was performed on these medians.

The RT for items in which only the first word has a same relation with the target did not differ from the RT for items in which only the second word has a same relation with the target: $F_1 < 1$. This means that no evidence was found for the left-to-right processing of the item. Accordingly, items that differ only with respect to the left-to-right order in the presentation are grouped together in the further analyses. Medians were computed for each of these groups for each subject. The means of these medians are presented in Table 4.9.

Table 4.9. Mean RT in ms for the item types in Experiment 4 (one target category)

Relation to target of Left-hand word	Right-hand word	RT
same	same	900
same	different-related	1106
same	different-unrelated	1020
different-unrelated	different-unrelated	900
different-related	different-related	993
different-related	different-unrelated	1038

The predictions were confirmed by the data. The first two hypotheses refer to the items in which one word is an exemplar of the target category. The RT for the same items was shorter than for the different items: $F_1(1,36) = 26.41$, $p < 0.001$; $min\ F'(1,45) = 15.01$, $p < 0.001$. The RT for the different-related items was longer than for the different-unrelated items: $F_1(1,36) = 5.51$, $p < 0.05$; $min\ F'(1,45) = 3.13$, $p < 0.05$ one tail.

The final predictions deal with the items in which both words have a different-unrelated relation to the target. The RT for these items was shorter than for the items in which a same word was combined with a different-unrelated word or with a different-related word: $F_1(1,90) = 17.28$, $p < 0.001$; $min\ F'(1,135) = 10.74$, $p < 0.001$. The former items differed also from the items in which a different-related word was paired with a different-unrelated word or with another different-related word: $F_1(1,90) = 11.68$, $p < 0.001$; $min\ F'(1,135) = 7.29$, $p < 0.01$.

4.4.3 Discussion

The present data are relevant for the model in several ways. The data on the three item types same same, same different-unrelated, and same different-related replicate the findings of the earlier experiments, but now on the task with only one target category.

No evidence was found for a left-to-right processing of the item in the one-target paradigm. A fortiori there is no ground for assuming left-to-right processing if there are more than one-target categories. The model in Fig. 4.1 is in accordance with this idea.

The data on the different different item types, e.g., *sparrow pigeon, sparrow tulip,* and *tulip rose*, can be incorporated in the model if it is extended with a preliminary stage. In that preliminary stage one checks whether or not the item bears any relation to the target category. If one or both of the words belong to the target category, the words are processed according to the original model and go through a verification and, if still appropriate, through a falsification stage. If none of the words belong to the target category, differences between the words and the target category have to be detected before the answer "different" can be given. It should be noted that this stage is entered only after a negative decision: none of the words belong to the target category. Therefore, this stage is the same as the falsification stage in the original model: the process in this stage starts with checking the difference on a higher order category level.

4.5 Conclusion

At several points in the model a matching process takes place. The nature of these matching processes has not been described in detail. The reason is that the successive steps in the model, which correspond to levels in a hierarchical organization, account for the major effects in the experiments. Relations other than hierarchical relations will no doubt play a role in effects on a more detailed level and in differences between individual words. Furthermore, one can think of tasks that do not appeal so heavily to the definitional and thus to the hierarchical meaning structure of the words in the present domain. Relations other than hierarchical relations may play an important role in those tasks. This is explored further in Part II.

Processes in semantic memory, as revealed in the present experiments, are determined by two basic factors. The first one is the verification/falsification order. When subjects have to judge whether there is a certain relation between concepts, they first try to prove that that relation exists. They do so by searching for positive evidence on the existence of the relation. If no positive evidence is found, a negative answer is required. This negative answer, however, is not given until positive evidence is found for the conclusion that the negative answer is correct. So, the second stage is a search for falsifying information.

The second basic factor by which the semantic processes are determined is the mental organization of the material as actualized by the nature of the task. Presumably the impact of the material is such that hierarchical processing factors are activated. The question is: when semantic judgments have to be made in a domain with a nonhierarchical underlying structure, do factors derived from the structure of that domain indeed guide the processing? This is the object of study in Part III on the domain of kinship terms.

An important characteristic of the model is its goal directedness. This is true at the very general level, as has just been discussed: in the process, evidence on similarity is searched for first and then evidence on difference. The goal directedness also occurs at a more detailed level. As soon as the category of a word has been retrieved, this category serves as the target. This target is going to control the process. It is especially in the one-target paradigm that this goal directedness is plausible. One might argue that the first word which is processed is the word whose category is the prespecified target. The fact that no evidence for the left-to-right processing has been found is in agreement with this goal directedness. This point is taken up again in the study on kinship terms in Part III.

Part II

Generalization to Another Task

Chapter 5 Judging the Semantic Similarity in a Hierarchical Domain

Processes in semantic memory, as studied in the previous part, turned out to be determined by two basic factors. The first factor is the verification/falsification order in the process. When subjects have to judge whether a certain relation between concepts exists, they first try to find positive evidence for that relation. If this cannot be found, they will try to find evidence that falsifies the relation. The second factor by which the semantic processes are determined is the mental organization of the material as actualized by the nature of the task. The material studied in the previous experiments was hierarchically ordered; the task required judgments with respect to these hierarchical structures. The verification and falsification processes have been characterized in terms of the hierarchical structure of the material.

The question to be studied in the present experiment is to what extent the model in the previous part can be generalized to other tasks as well. In the previous tasks, subjects were instructed to judge the similarity between concepts with respect to their hierarchically related categories. The task induced a hierarchical way of processing and the RTs could indeed be accounted for by the hierarchical relations between the concepts.

The material used in the present experiment is the same as in the previous part but the task differs. In the previous experiments, concepts were judged with respect to prespecified categories. In the present experiment, subjects were required to judge semantic similarity between the concepts. No dimensions for similarity were specified. On the contrary, subjects had to indicate their subjective feeling about the semantic similarity. In order to allow subjects to differentiate their judgments they were required to give their similarity judgments on a seven-point scale. The scale was labeled from 1 (no similarity) to 7 (great similarity). So, the present task does not emphasize the hierarchical structure of the domain but allows for a variety of factors in determining the semantic similarity. Both the scale value, attributed to a word pair, and the RT for the judgment were recorded.

If the principles of the model have validity for the present task, the following results are to be expected. First, there will be a process of judging the similarity between concepts as distinct from a process of judging the difference between concepts. Second, subjects will ask the question about similarity before evaluating the difference between concepts. Third, if the hierarchical model has some generality, the judgmental process will at least partly be determined by the hierarchical classification of the word pairs in same, different-related, and different-unrelated.

Given the objective of collecting judgments on a seven-point scale, some methods were considered for the execution of the responses. One could construct a scale consisting of seven response buttons in a row corresponding to the values one to seven. But the position of the response button in the row has an effect on the RTs. An experiment was performed [5.1] in which subjects were presented with the digits from 1 to 7. They were required to press a corresponding response button in a row of seven labeled buttons. The means shown in Fig. 5.1 indicate that the RT will be lower the more extreme the position is on the scale. Consequently, when analyzing the RTs for the semantic judgments, one ought to take the effect of the response buttons into account. One procedure is to subtract the RT for the execution of the answers from the RT for the semantic judgments. This is the procedure used by CARAMAZZA et al. [5.2]. It is questionable, however, whether this subtraction procedure is correct. Results in experiments of NOORDMAN-VONK [5.3] cast some doubt on the independence in RT of the process of response execution and of stimulus evaluation.

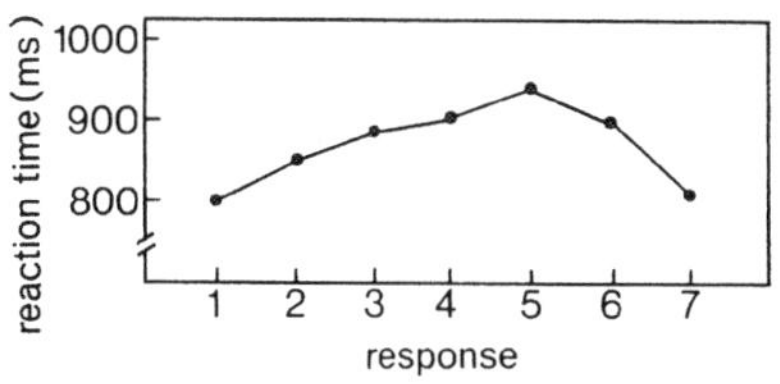

Fig. 5.1. Mean RT in ms for the responses on a scale of response buttons to the digits 1 to 7

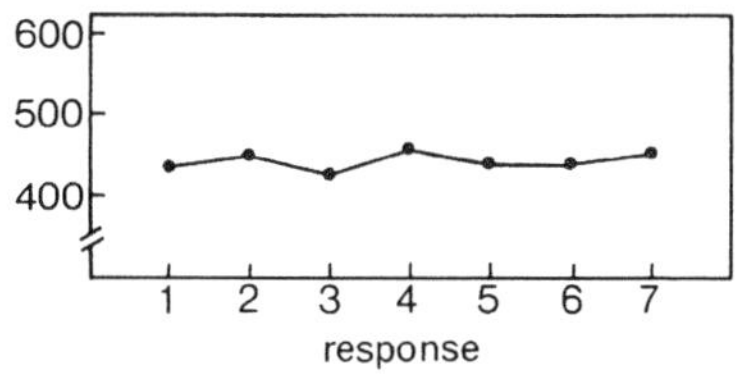

Fig. 5.2. Mean RT in ms for the spoken responses (voice key) to the digits 1 to 7

Another way to obtain judgments on a seven-point scale is using a voice key. The experiment in which subjects were presented with digits from 1 to 7 was also performed with a voice key as the response medium. The means shown in Fig. 5.2 indicate that the differential effect of the voice key response execution on the RTs is much smaller and that RT is not a function of the (central or extreme) position on the scale. Therefore, the method of collecting data on semantic judgments by means of a voice key seems indicated.

The question, however, is whether subjects are able to make the same sort of judgments in the case of voice key response execution when there is no physical response scale, as in the case of response execution on a response button panel. One might hypothesize that subjects, when the physical scale is not available, will not fully exploit the possibilities to give differentiated answers on the scale, or that they are more liable to uncertainty, which might result in low reliability.

In the present experiment on judging semantic similarity, therefore, judgments were made using a response button panel as well as a voice key. One group of subjects (group A) had to execute the responses in the practice blocks with a scale of response buttons and in the experimental blocks with a voice key. In order to investigate whether the numerical value of the responses with the voice key were affected by the preceding trials with the response button panel, the other group of subjects (group B) had to give vocally the responses in the practice blocks as well as in the experimental blocks.

5.1 Method

Material. The same material as in the previous experiments was used: all words were exemplars of the categories *tree, flower, bird,* and *insect*. The exemplars are presented in Table 3.2. All words occurred an approximately equal number of times in the experiment. Because of the saliency of the level of the categories (*tree, flower, bird,* and *insect*) (cf Experiments 2,3, and 4 of Chap. 3), word pairs were categorized with respect to these categories in same, different-related, and different-unrelated item types (e.g. *oak beech, oak rose,* and *oak sparrow*, respectively; see Table 3.4).

There were 48 different experimental items, 16 items for each of the item types. The 48 items were presented twice in four blocks of 24 items each. There were two different orders of the items in the blocks. The blocks were presented in four different orders. Each block contained eight same word pairs, two from each of the four categories; eight different-related word pairs, four from each of the two kinds of different-related word pairs; and eight different-unrelated word pairs, two from each of the four kinds of different-unrelated word pairs. Two practice items were added to each block. Not more than two consecutive items were of the same type. No word occurred more than once in three successive items. All words occurred an approximately equal number of times in the experiment. The words were presented in a left-to-right order arrangement. Each word occurred an approximately equal number of times in the left-hand position and in the right-hand position.

Procedure. Four practice blocks preceded the experimental blocks. Twenty subjects (group A) gave the responses in the practice blocks by means of a scale of response buttons; fifteen subjects (group B) gave the responses in the practice blocks by means of a voice key. The practice blocks were identical to the experimental blocks.

All the words that were used in the experiment were shown to the subject on a card before the experiment. The subjects were instructed to rate the semantic similarity between the words on the basis of their immediate and overall feeling. They had to express the degree of similarity in meaning with the digits 1 (no similarity in meaning) to 7 (very great similarity in meaning). The subjects had to give their responses as rapidly as possible.

The items were presented with three slide projectors. The distance of the subject to the screen was approximately 2.1 m. The items subtended a visual angle of about 12 deg in a horizontal direction. During a period of 1.5 s before the presentation of an item, a fixation point was presented. After the answer was given, there was a blank interval of 3 s. The RT was measured from the onset of the presentation of the item until the subject gave his answer, and was recorded on digital tape. The response was recorded on audio tape.

Subjects. Thirty-five undergraduate students served as subjects; the majority of them were students in psychology and biology. They were paid for their participation.

5.2 Results

In order to ascertain whether subjects make the same sort of judgments with a voice key as with a response button panel, the judgments in the second part of the practice trials (button panel) and in the first part of the experimental trials (voice key) were compared for group A. For each subject the number of different scale values given when using a voice key and when using a response button panel were determined. The number of scale values given in the voice key answering mode did not differ from the number given in the response button panel mode: Wilcoxon matched-pairs signed-ranks test $p > 0.25$ ($z = 1.43$). Furthermore, for each of the subjects the correlation between the scale values given with the response buttons and given with the voice key was computed over the items. The average product moment correlation, computed with Fisher's z transformation, was 0.86. It can be concluded that subjects are able to evaluate semantic similarity on a seven-point scale equally well when a physical scale is absent as when such a scale is available.

In order to determine whether the judgments on similarity with the voice key were affected by the preceding responses with the button panel, the judgments in the experimental blocks of group A (practice trials with response buttons) and group B (practice trials with voice key) were compared with respect to the number of scale values subjects used and with respect to the reliability of the judgments.

For each subject the number of different scale values used was determined. The distribution of subjects over the possible number of scale values for group A was not different from that distribution for group B: Kolmogorov-Smirnov two sample test $\chi^2(2) = 3.43$, $p \gg 0.20$ [Ref. 5.4, p.131]. Thus, the number of different scale values used by the subjects did not differ for the two groups. The number of scale values that were used, averaged over all subjects, was 6.0. Only two subjects used as few as 3 scale values. The percentages of the responses for the scale values 1 to 7, averaged over subjects, was 27, 13, 9, 10, 14, 13, and 14, respectively. So the subjects used the scale in a very differentiated way when expressing the semantic similarity.

It should be kept in mind that the 96 experimental items consisted of 48 different items, presented in two different orders. For each subject the correlation between the one half and the other half of the items has been computed. The average product moment correlation, computed with Fisher's z transformation, was 0.93 for group A and 0.89 for group B. So, the groups were also very similar with respect to the reliability of the judgments.

It can be concluded that group A (practice trials with response buttons) and group B (practice trials with voice key) do not differ with respect to the similarity judgments given in the experimental blocks. Therefore, the two groups will be treated as one group in the following analyses.

The reliability of the judgments, as expressed by the product moment correlation, averaged over all subjects was 0.91. Parenthetically, the high correlation between the judgments made with voice key and with response buttons had already testified to the reliability of the judgments. Inspecting the Pearson correlations of the individual subjects, it appeared that five subjects had a strikingly low reliability as compared with the other subjects. They averaged 0.57, the highest of them being 0.64, while the average of the other 30 subjects was as high as 0.94, the lowest being 0.77. Because of their low reliability, the results of these five subjects have not been pooled with those of the other subjects in the following analyses.

The important point with respect to the similarity values is the correlation with the a priori classification of the items in same word pairs, different-related word pairs, and different-unrelated word pairs. A score was attributed to each word pair, expressing the semantic similarity between the words: 3 for same, 2 for different-related, and 1 for different-unrelated. For each subject the correlation between these a priori scores and the semantic judgments was computed. The average product moment correlation, computed with Fisher's z transformation, was 0.84.

Another way to evaluate the correspondence between the semantic judgments and the hierarchical classification of the material is to compute the mean scores for the same, different-related, and different-unrelated word pairs. These results, presented in Table 5.1, indicate a high correspondence between the a priori classification and the subjective judgments.

Table 5.1. Mean and standard deviation of the similarity judgments for the item types same, different-related, and different-unrelated

Item type	Mean	Standard deviation
same	5.79	0.98
different-related	3.44	0.48
different-unrelated	1.95	0.30

Before proceeding to the results on the RTs, a few remarks should be made with respect to the statistical tests. The tests were again performed on median RTs computed per subject: the median of the RTs in a particular response category, e.g., the response value 5, or in a particular item type-response combination, e.g., the response value 5 for same word pairs. Not all response values were used equally frequently by all subjects. The general rule that was adopted in computing medians per subject was that no median was computed for a response category or an item type-response combination if the subject had six or less RTs in that category or combination. In such a case, the observations were not included. Moreover, item type-response combinations with a median for only one subject were not included, either. There were three of such item type-response combinations with one subject, and the next smallest number of subjects in an item type-response combination was six.

Because not all subjects are represented in every mean, comparisons between means over all data regard intersecting sets of subjects. Consequently, the observations are neither completely dependent nor completely independent. Therefore, the following procedure was used. In comparisons between two mean RTs, only those subjects were considered who were represented in both means. Thus, all analyses were based on matched observations.

Another reason for testing correlated samples only is the fact that the meaning of the scale values may very well be different for different subjects. For example, the scale value 4 for one subject might be equivalent to the scale value 5 for another subject. Therefore too, only matched observations were analyzed in comparing the RTs for response categories and for item type-response combinations. So, the scale values are treated as ordinal scores.

To what extent do the RTs for the judgments depend on the semantic similarity as judged by the subjects? For each subject the median RT was computed for each response category. The means of these medians are presented in Fig. 5.3. The RT is the shortest for the most extreme values on the scale. The more similar the words are judged, the shorter the RT is, and the more different the words are judged, the shorter the RT is. The longest RT is obtained for intermediate degrees of similarity. These results agree with most of the models in semantic memory, as was discussed in the Introduction.

A more specific point with respect to the dependency of the RTs on the semantic similarity is whether subjects ask the question about similarity before evaluating the semantic difference between concepts. This point is, of course, related to the same/different effect in the verification experiments (Chap. 4). The RTs for the responses very similar (7) and not similar at all (1) were compared. The RT of the responses very similar is 219 ms shorter than the RT of the responses not similar at all. When considering only matched observations, this difference is 253 ms; $t(19) = 4.31$; $p < 0.001$. The conclusion is that the data agree with a model in which evidence on sameness is first searched for.

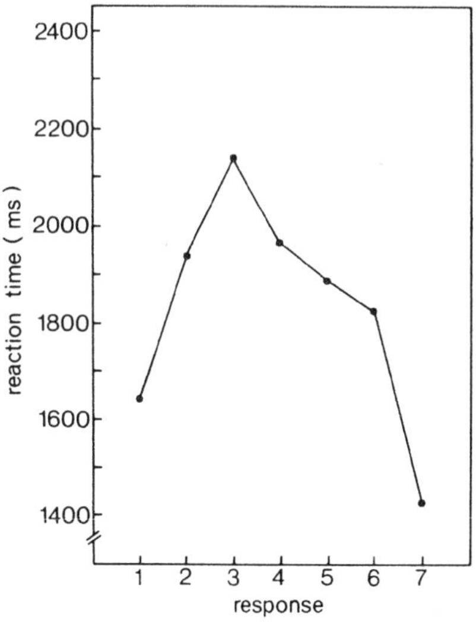

Fig. 5.3. Mean RT in ms for the similarity judgments over all item types

As was made clear earlier, the present task does not emphasize the hierarchical structure of the domain but allows for a variety of factors in determining the semantic similarity. The question is to what extent the RT depends on the semantic similarity as defined by the hierarchical relations in the material or by other features. For each subject the median RT was computed for the item type-response combinations. The means of these medians are presented in Fig. 5.4.

The RT for same word pairs increases the less similar the words are judged. All these items are identical with respect to the hierarchical structure. So, the semantic similarity as defined with respect to the a priori hierarchical relations is interfered with by other factors on which the word pairs differ. The analogous result was found for the different-unrelated items. The less different the words are judged, the longer the RT is. Again, this difference must be determined by factors other than the hierarchical relations in the material, because in terms of the hierarchical relations all these items are different-unrelated items. Parenthetically, these results suggest that the different-related items might be subdivided into two groups depending on the degree of similarity. The RT for the more similar group of different-related items increases the less similar the words are judged. The responses for the more different group of items are delayed the more similar the words are judged.

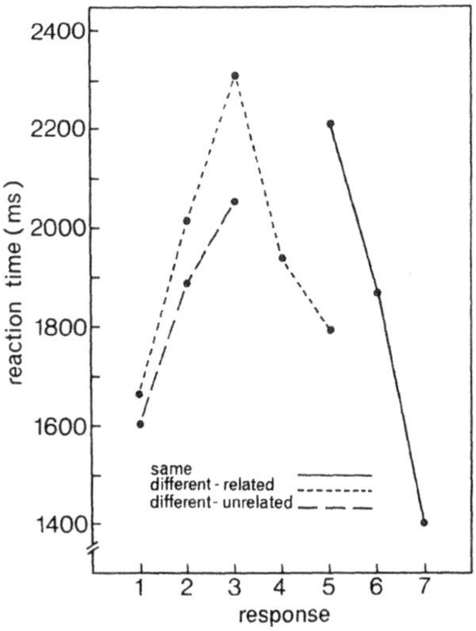

Fig. 5.4. Mean RT in ms for the similarity judgments for the item types separately

However, the RT is not fully determined by the subjectively evaluated similarity. The RTs for items that are judged as equally similar are not equal but vary systematically with the a priori classification of the items. The RT for same word pairs that are judged equally similar as different-related word pairs is longer than for the corresponding different-related word pairs. So, the conflict between the similarity in terms of the hierarchical relations and the relative dissimilarity in terms of other features slows down the response. The RT for different-unrelated word pairs that are judged equally different as different-related word pairs is shorter than for the corresponding different-related items. In this case the conflict between the difference in terms of hierarchical structure and the relative similarity in terms of other features slows down the response. All the observed differences in RT between word pairs that are judged equally similar correspond to this pattern: the RT for different-unrelated items that are judged 1 is 183 ms shorter than the RT for different-related items that are judged 1: $t(10) = 2.21$, $p < 0.05$ one tail. The differences for the judged similarity 2 and 3 are, respectively, 171 and 102 ms: $t(4) = 1.83$, $p < 0.10$ one tail and $t(5) = 1.66$, $p < 0.10$ one tail, respectively. The RT for same items that are judged as 5 is 247 ms longer than the RT for different-related items that are judged equally similar: $t(3) = 1.91$, $p < 0.10$ one tail. Each of the four comparisons illustrates the role of the semantic similarity as defined by the hierarchical structure in determining the RT. The number of subjects involved in these tests was very small: 11, 5, 6, and 4 subjects, respectively. There is not much overlap between these groups: the 26 differences are derived from 20 different subjects. The effect of the semantic similarity, as defined by the hierarchical structure, was tested for these 20 subjects. The differences in RT between the a priori hierarchical classification for the four response values on the similarity scale were significant: $t(19) = 3.09$, $p < 0.01$. It is clear that the RT not only depends on the overall judged semantic similarity between the concepts but is determined substantially by the hierarchical relations in the material.

5.3 Discussion

On the basis of the results, the process of judging the semantic similarity may be described in the following way. The first step in the comparison process is the search for similarity. In case the words are very similar with respect to both the hierarchical relations and other properties, the answer "7" is executed immediately. In other cases the degree of similarity has to be evaluated, and the similarity in terms of hierarchical relations is interfered with by semantic difference due to other factors. If, at the start of the process, the words are not judged as similar with respect to the hierarchical relations, an evaluative process for differences takes place. If the words are very different with respect to both the hierarchical relations and other properties, the answer "1" is given. Else, the degree of difference has to be evaluated. The less different the not-similar words are judged, the more the answer will be delayed: the difference in terms of the hierarchical relations is interfered with by the semantic similarity in other factors. The order of the search for similarity and for difference accounts for the fact the judgments very similar (7) are produced faster than the judgments very different (1).

It should be clear that this process of judging the semantic similarity resembles in all respects the model developed in the previous part. There appears to be a process of judging semantic similarity as distinct from the process of judging semantic difference. The nature of these processes is determined to a large degree by the hierarchical structure of the material in spite of the fact that the task did not stress this hierarchical structure. Semantic similarity is judged at an earlier stage of the process than is semantic difference.

Part III

Generalization to Another Semantic Domain

Chapter 6 The Kinship Domain

The retrieval processes in the previous experiments were determined by two factors: the set for verification and the structure of the domain as actualized by the task and the criterion. The aim of the present part is to investigate to what extent the principles of the previous model can be generalized to another, nonhierarchically organized, domain. To study this, verification experiments were performed in a paradigmatic field, viz., the kinship domain, a clearly nonhierarchically organized domain as was described in the Introduction. The verification experiments, reported in Chap. 7, are similar to those performed on the hierarchical domain, reported in Chap. 4. The word pairs are formed with kin terms from the nuclear family. The subjects have to judge whether both words are of the same sex or not, or whether the two words express a parent-child relation or not. The present chapter deals with the role of the structure of the kinship domain in the verification process.

The structure of the kinship domain has been studied extensively, and several componential systems have been proposed [6.1-3]. In these systems three or more of the following components play a role: sex, generation (or *parent of* and *child of* relations), lineality, and reciprocity. The psychological relevance of the systems has been tested in [Ref. 1.12, Chap. 4; 6.2,4,5]. The componential system of ROMNEY and D'ANDRADE [6.2] seems to reflect the cognitive structure of the domain rather well. The system does, however, not specify a difference in importance of the features in characterizing the cognitive structure. The findings on similarity judgments in the study of WECHSLER and ROMNEY [6.4] suggest this difference in importance of the features. FILLENBAUM and RAPOPORT [1.12], in testing the componential systems, presented the cognitive space for the kinship terms by means of two-dimensional representations.

The question is what predictions can be made from these studies with respect to the retrieval processes in verification experiments. One may assume that cognitive similarity is important in determining the retrieval order of the concepts: the more related the concepts are, the closer to each other they will be in the retrieval order. For example, the results of the aforementioned scaling studies will predict that terms that differ only in sex will be retrieved shortly after one another. But the structure of the domain, as reflected in representations obtained from scaling studies, provides at best indirect predictions for the retrieval processes. They do not indicate the saliency of the terms. This saliency, though, was indicated by the data, collected in testing the psychological relevance of the systems. ROMNEY and D'ANDRADE found differences in the retrieval order of the terms in an experiment in which they had subjects list names for kinds of relatives and family members. The saliency of the terms was also illustrated by the results of graph and cluster analyses in the studies of FILLENBAUM and RAPOPORT. The saliency of the kin terms is likely to be an important factor in determining the retrieval order of the terms. Accordingly, the saliency of the terms is predicted to be an important factor in determining the processes in the verification tasks on the kinship domain.

A qualification with respect to this expectation should be made. If in the verification task the two words differ on the feature that is the criterion in the task, it may very well turn out that the words are not retrieved in order of their saliency. Indeed, because the set for verification controls a search in function of the criterion, words that are the same with respect to the criterion will be searched for first. The set for verification and the role of the feature that is the criterion in the task will then overrule the role of the structure of the domain. Accordingly, the RT for the judgment whether *brother* and *sister* are of the same sex might very well differ from the RT for the judgment of whether there is a parent-child relation between *brother* and *sister*. The point is not so much that the absolute size of the RTs will differ, but that the relative RT of an item as compared to the other items will vary depending on the criterion.

In the previous chapter it was concluded that the judgment processes in word comparisons are determined substantially by the structure of the semantic domain as actualized in the task and not only by the overall judged semantic similarity between the concepts. In order to be able to investigate the generality of this conclusion, two experiments were performed on the organization of the kinship domain. Data were obtained on the judged semantic similarity between kinship terms (Sect. 6.1) and on the saliency in retrieval of kinship terms (Sect. 6.2).

6.1 Experiment 1: Judging the Semantic Similarity Between Kinship Terms

Kinship terms of the nuclear family were studied in this experiment: *father, mother, son, daughter, brother,* and *sister*. These are the terms that were used in the verification experiments, reported in Chap. 7. In a triadic comparison task all possible triples of these terms were judged. Each triple was used in the six possible configurations. The experimental items were presented in six blocks of the 20 different triples each. The word triples were distributed equally among all parts of the blocks; the sum of the rank-orders for the positions of the triples in the blocks were approximately equal. The sequence of items was such that no word occurred in more than two successive items and that no word pair occurred more than once in three successive items. The experiment was performed with a PDP-8E computer. The presentation of items on an oscilloscope and the recording of responses were under program control. A subject had to indicate for each triple of words which two words were most similar in meaning by touching the line on the response panel that corresponded to the side of the imaginary triangle connecting the two words. The subjects were 20 students, the majority of them undergraduates in psychology or biology. They were paid for their participation. One-half of the subjects were male, one-half female.

Results. Table 6.1 presents the percentage each particular word pair was judged as most similar in meaning in the triples of words. Word pairs consisting of words with more components in common were rated as the more similar, both by males and females: *father son, mother daughter* as opposed to *father daughter, mother son; father mother, son daughter, brother sister* as opposed to all other nonlinear word pairs. These data can be used to determine the semantic relatedness of the items in the verification experiments of Chap. 7.

6.2 Experiment 2: The Production of Kinship Terms

This experiment replicates the free recall listing experiment of ROMNEY and D'ANDRADE [6.2] for a Dutch sample of subjects. The task required subjects to name kinship terms in the order in which they came to mind. In the instruction subjects were told: "There are words that indicate family members and relatives. Will you please name them?" The answers of the subjects were recorded on tape. The subjects were 100 students, the majority of them undergraduates in psychology. Half of them were male, the other half female. They volunteered in the experiment.

Table 6.1. Mean percentage of "most similar" judgments, for all subjects (T), and for male (M) and female (F) subjects separately

Word pair		T	M	F
1) father	son	11	10	12
2) father	daughter	6	4	8
3) mother	daughter	11	9	12
4) mother	son	6	4	8
5) father	mother	17	17	16
6) son	daughter	13	14	11
7) brother	sister	15	15	15
8) father	brother	3	3	3
9) father	sister	0	0	0
10) mother	sister	4	4	3
11) mother	brother	0	0	0
12) son	brother	5	6	4
13) son	sister	2	3	1
14) daughter	sister	5	6	4
15) daughter	brother	3	4	2

Results. The total number of terms that were named was 1313; accordingly, the subjects produced on the average 13 kin terms. Both production frequency and mean position in the naming sequence are taken as indices of saliency for the different kinship terms. Table 6.2 presents the kin terms with a production frequency of at least 15. They are rankordered according to mean position in the naming sequence. For each term the production frequency and the mean position in the naming sequence are presented, computed over all subjects as well as over male and female subjects separately.

From Table 6.2 it is clear that the terms differ in saliency. There are terms that are named by almost all subjects: *father, mother, uncle*, and *aunt* are named by over 90% of the subjects; *brother, sister, nephew* or *cousin, niece* or *cousin* (the Dutch *neef* means *nephew* and *cousin*, and the Dutch *nicht* means *niece* and *cousin*), *grandfather* and *grandmother* by 81-90% of the subjects. The production frequency drops to less than 50% for the other terms: *grand nephew* or *second cousin, grand niece* or *second cousin*; and, surprisingly, the other two terms of the nuclear family *son* and *daughter* are named by only one-third of the subjects. The fact that *father* and *mother* are so much

Table 6.2. Production frequency and mean position in naming sequence of terms for family members and relatives, for all subjects (T), and for male (M) and female (F) subjects separately

	Production frequency			Mean position in naming sequence		
	f[a] T	% M	% F	T	M	F
vader *father*	93	90	96	3.41	3.96	2.90
moeder *mother*	91	90	92	4.04	4.51	3.59
oom *uncle*	95	96	94	4.72	4.48	4.96
broer *brother*	87	88	86	4.87	5.91	3.81
tante *aunt*	96	96	96	5.43	4.90	5.96
zus *sister*	87	88	86	5.59	6.66	4.49
neef *nephew/cousin*	88	90	86	5.70	5.04	6.40
nicht *niece/cousin*	83	84	82	7.04	6.93	7.15
zoon *son*	34	44	24	8.24	7.64	9.33
opa *grandfather*	84	78	90	8.37	8.36	8.40
dochter *daughter*	35	44	26	8.54	8.18	9.15
kind *child*	17	16	18	8.59	8.63	8.56
oma *grandmother*	85	80	90	8.91	9.23	8.62
achterneef *grand-nephew/second cousin*	42	42	42	12.40	11.57	13.24
kleinkind *grandchild*	21	20	22	12.48	12.00	12.91
achternicht *grand-niece/second cousin*	42	42	42	13.02	12.62	13.43
zwager *brother-in-law*	28	30	26	13.11	12.60	13.69
oudtante *great-aunt*	15	12	18	13.73	13.50	13.89
schoonzus *sister-in-law*	18	12	24	14.22	15.33	13.67
overgrootvader *great-grandfather*	19	22	16	14.58	14.82	14.25
overgrootmoeder *great-grandmother*	17	20	14	15.29	15.50	15.00

[a] n = 100

Note: If a particular relative was denoted in the experiment by several synonymous terms (e.g., *opa*, *grootvader*), the frequencies were summed up but only the dominant term is presented in the table.

more salient than *son* and *daughter* for these subjects may be due to the age of the subjects and their place and orientation in their own family [Ref. 6.6, p.74]. The mean position in naming sequence gives results similar to the production frequency. Accordingly, there is a high correlation between both indices: r = -0.86. There are, however, striking differences: *grandfather* and *grandmother* have high production frequencies, but are relatively late in the naming sequence; and conversely, *child* is named infrequently and when it is, it is named relatively early in the naming sequence.

Terms that differ only in the feature sex are named immediately after each other, the male term first, then the female term, for both male and female subjects. There appears to be, in fact, no difference whatever in the results for male and female subjects.

The results correspond to the results of the listing task of ROMNEY and D'ANDRADE. As has been pointed out, only terms of the nuclear family were used in the verification experiments, reported in the following chapter. These terms differ in saliency. Parent terms are more salient than sibling terms, child terms are less salient. It is assumed that the saliency of the terms is related to the order of processing the terms. Therefore, the results of the experiment give clear predictions with respect to the retrieval process in verification experiments on the terms of the nuclear family. On the basis of the structure of the domain, the order of processing is predicted to be parents (*father* and *mother*) first, siblings (*brother* and *sister*) second, and finally, children (*son* and *daughter*). As was argued earlier, however, the set for verification and the role of the feature that is the criterion may overrule the role of the structure of the domain as expressed in the saliency of the terms.

Chapter 7 The Verification of Kinship Relations

The aim of this chapter is to study the generality of the principles of the model, developed for the verification of hierarchical relations. The question is to what extent the RTs for semantic judgments on kinship terms are determined by the verification/falsification order and by the actualized structure of the domain. Terms of the nuclear family were used in the present experiments: *father, mother, son, daughter, brother,* and *sister*. On each trial in the experiments two kinship terms were presented to a subject. He had to judge these terms with respect to a specified criterion. In the first experiment the criterion is sex: the subject had to judge whether the two presented concepts were of the same sex. The criterion in the second experiment is the parent-child relation: the subject had to judge whether the two terms did express a parent-child relation.

The semantic similarity between concepts in the hierarchical domain is intrinsically related to the criterion, e.g., the similarity between *oak* and *beech* as well as between *oak* and *tulip* is related to the criterion *tree*. Semantic similarity conceived of in this way is not involved in the judgments of kinship terms with respect to the criteria sex or parent-child relation. Semantic similarity can also be defined, however, in terms of feature overlap, independent of any criterion. The question is whether semantic similarity conceived of in this way plays a role in the judgments of kinship terms. When the criterion is sex, semantic similarity may be defined in terms of, for example, parent-child relation, generation, and lineality. When the criterion is parent-child relation, semantic similarity can be defined in terms of sex, generation, and lineality. This, however, leads to somewhat arbitrary categorizations for the items that require the answer "different". When the criterion is sex, and if one emphasizes the feature parent-child relation, items such as *father mother* and *father daughter* will be called different-unrelated and different-related items, respectively, but they will be called different-related and different-unrelated items, respectively, if one emphasizes the feature generation. Similarly, when the criterion is parent-child relation, *father mother* and *father brother* will be called different-related and different-unrelated items, respectively, if one emphasizes the feature generation, but they will be called different-unrelated and different-related items, respectively, if one emphasizes the feature sex. Therefore, if one defines relatedness with respect to these a priori features, the semantic distance effect is investigated for same items only, where such arbitrary categorizations do not occur. Relatedness, however, can be defined with respect to rated similarity as well. Data on rated similarity were obtained in Experiment 1 of Chap. 6 (Table 6.1). Is semantic similarity, conceived of as rated similarity or as similarity in terms of a priori features, expected to play a role in the judgments? Semantic similarity in the hierarchical domain was conceived of only with respect to the criterion and not with respect to any features whatsoever of the concepts. If, in line with the model developed in Part I, the judgments depend mainly on semantic similarity with respect to the criterion, one expects that the similarity effect plays at most a minor role in the judgments with respect to sex and parent-child relation.

As was indicated above, the retrieval of semantic information on the kinship domain was investigated in two experiments, in which subjects had to judge whether two terms are of the same sex (Experiment 1) and whether they express a parent-child relation (Experiment 2).

7.1 Experiment 1

The criterion in the present experiment was sex: subjects had to judge whether two presented words are of the same sex.

It was hypothesized that subjects will first try to find evidence confirming the conclusion that the two terms are of the same sex. If this evidence cannot be found, they will try to falsify the criterion, i.e., to prove that the conclusion "false" is correct. Consequently, the RT for same word pairs was expected to be shorter than the RT for different word pairs.

A second prediction is related to the structure of the kinship domain. From the results of the naming task experiment, reported in Chap. 6 (Table 6.2), the order of processing was predicted to be parents first, siblings second, and lastly, children. Consequently, it was expected that the RT would be shorter for items with the term *father* or *mother* than for items with the term *son* or *daughter*, and that the RT for items with the term *brother* or *sister* would be in between. Furthermore, it was predicted that there would be no similarity effect.

Table 7.1. Mean RT in ms for judgments with respect to the criterion same sex (Experiment 1)

Item	Sex of subjects Male	Female
father son /son father	833	932
father daughter/daughter father	976	1019
mother daughter/daughter mother	911	922
mother son /son mother	950	906
father mother /mother father	946	926
son daughter /daughter son	1059	1017
brother sister /sister brother	991	1024
father brother /brother father	887	917
mother sister /sister mother	917	847
son brother /brother son	883	1007
daughter sister/sister daughter	1007	928

7.1.1 Method

Material. The experimental material consisted of 22 word pairs of the six kin terms *father, mother, son, daughter, brother,* and *sister*. The items can be considered as 11 item types, if the left-right order is not taken into account. The items are presented in Table 7.1. Other combinations of the six kinship terms served as practice items. The items were presented in blocks of 24 items each. There were four experimental blocks, preceded by two practice blocks. There were two practice items at the beginning of each block. The word pairs were distributed equally among all parts of the blocks; the sum of the rank orders for the positions of the word pairs in the blocks were approximately equal. The number of same items was approximately the same as the number of different items. The sequence of items was such that not more than three consecutive items required the same answer, and that no word occurred more than once in two successive items.

Procedure. The experiment was performed with a PDP-8E computer. The items were presented on a cathode-ray oscilloscope. The timing and the recording of responses were under program control. The distance of the subjects from the display was approximately 0.65 m. The items subtended a visual angle of about 10 deg in a horizontal direction. During a period of 0.75 s before the presentation of an item, a fixation point was presented. After the answer was given, there was a blank interval of 2 s. The RT was measured from the onset of the presentation of the item until the subject gave his answer. The subject had to press a "same" or "different" button with his index finger to indicate the answer. For one-half of the subjects the "same" button had to be pressed with the right index finger; for the other subjects the "same" button was assigned to the left index finger.

Subjects. The subjects were 20 students, the majority of them undergraduates in psychology or biology. They were paid for their participation. One-half of the subjects were male, one-half female.

7.1.2 Results

The number of errors was on the average 1.9%. No subject made more than 5% errors. For each subject, 22 medians were computed for the correct responses, one for each item. A three-way analysis of variance was performed on these medians. There were two between subjects factors: sex of subjects and assignment of the response buttons to the index fingers. The within subjects factor had 22 levels, viz., the items.

The predictions were confirmed by the data. The RT for the same items was 66 ms shorter than for the different items: $F(1,336) = 36.49$, $p < 0.001$. The RT for items with at least one parent term was 77 ms shorter than for items with at least one child term: $F(1,336) = 24.60$, $p < 0.001$. That the RT for items with a sibling term is in between items with a parent term and items with a child term is indicated by the RT for the items *father mother/mother father*, *brother sister/sister brother*, and *son daughter/daughter son*: these were 936, 1007, and 1038 ms, respectively.

Some other effects were found. The sex of the subjects interacts with the items [$F(21,336) = 2.09$, $p < 0.01$, although $p > 0.10$ if the degrees of freedom were set at the lower bound in the Geisser-Greenhouse procedure], in particular with the items in which a parent term occurred. The RT is 91 ms shorter if sex of subject and parent term is the same than when the sexes differ: $F(1,336) = 8.63$, $p < 0.01$.

The 22 items constituted 11 item types if the left-right presentation of the two kinship terms is not taken into account. The order of the kinship terms had no effect on the RT. The difference in RT between two items in an item type did not reach a significance level of 5%, except for the item type *father brother* for female subjects ($p < 0.05$). The 22 items will be treated as 11 item types. The RTs in Table 7.1 are averaged over the two items in each type, separated for males and females.

The semantic similarity between the kin terms had no effect on the RTs. Whether one defines same-related and same-unrelated items with respect to one or several of the features generation, parent-child relation, or lineality, the difference between related and unrelated items was only 24 ms: $F(1,336) = 2.13$, $p > 0.10$. If one defines the related and unrelated items with respect to judged similarity data (Table 6.1), the RT for the same-related items is again these 24 ms shorter than the RT for same-unrelated items, and the RT for different-related items is 31 ms longer than the RT for different-unrelated items [$F(1,336) = 3.21$, $p > 0.05$]. The product moment correlation between similarity judgments and RTs for the same word pairs was -0.21 and for the different word pairs 0.13.

In order to test the reliability of the data, a split-half correlation was computed on item types, by dividing the subjects into two groups, which are matched with respect to sex of the subjects and left-right key assignment. The average product moment correlation was 0.86 (for male subjects 0.89 and for female subjects 0.83).

7.1.3 Discussion

The data give evidence for some factors underlying the judgment processes. First, subjects try to find positive evidence with respect to the criterion; if this evidence cannot be found, a falsification process follows. Second, there appears to be an order of searching information: parent terms first, then sibling terms, and then child terms. Third, subjects process kinship terms (at least parent terms) of their own sex earlier than kinship terms (parent terms) of the opposite sex, as indicated by the interaction of the sex of the subjects with the items. The second and third factors can be interpreted as the personal orientation of the subject. These factors determine the process, which is described by the process model in Fig. 7.1.

The first step in the process is to investigate whether the item contains the parent of the subject's own sex. If not, the subject is going to search for the sibling term of his own sex. If this term is not found, he will finally look for a child term of his own sex. If none of these questions is answered positively, the subject is going to search for a parent, sibling, or child term, respectively, of the opposite sex. If any of these questions is answered positively, the same questions are asked with respect to the other term in the item. The order of these questions is the same: parent, sibling, child. The sex of the other kinship term to be searched for corresponds to the sex of the term that has already been detected.

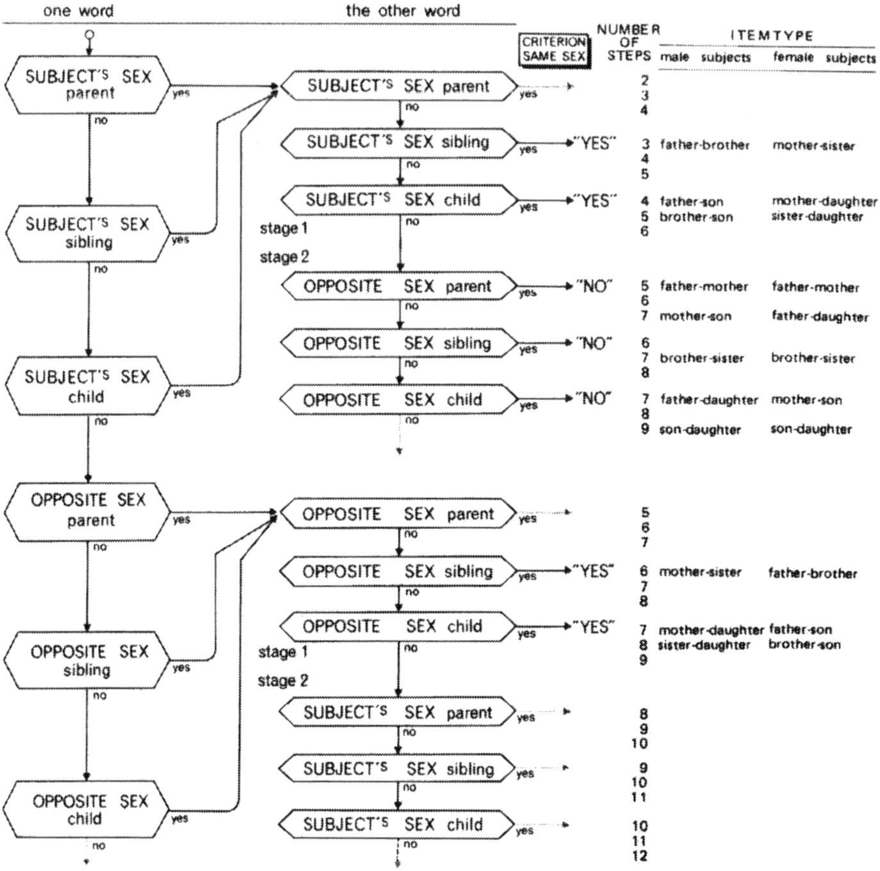

Fig. 7.1. Process model for the verification of kinship terms with respect to the criterion sex in Experiment 1

A few remarks should be made with respect to the model. It accounts for more items than were included in the experiment. In this way the model can be tested in further experiments. One restriction should be made, however: it seems very unlikely that the model will account for the RTs of items consisting of two identical kinship terms. Some global check will be needed in that case that,e.g., determines the physical similarity of the two terms.

Given the specifications of the model, the effect of the left-right order of presentation of the terms on the RT can be studied in greater detail. The number of boxes that must be passed in the judgment of an item will be called the number of steps of the item. Suppose the model is a description of a process that takes the information from left to right. The number of steps for each item can then be determined. This number varies from 3 to 12. Items in each item type were compared with respect to the number of steps. Items in five types differ in the number of steps. For these item types, the items with the larger number of steps, and the items with the smaller number of steps, averaged over item types, did not differ in RT (the RT difference was in fact 6 ms in the wrong direction). This leads to the conclusion that the process is indeed not a left-to-right process.

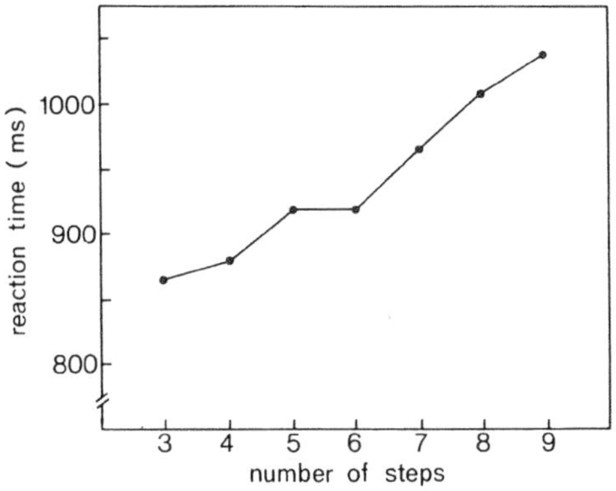

Fig. 7.2. The fit of the model to the results in Experiment 1

To what extent does the model account for the data? For each one of the 11 item types both for male and female subjects the number of steps in the model was determined. This number varies from 3 to 9. The fit of the model can be expressed by means of a correlation of the number of steps with the RTs. The Pearson correlation is 0.89. For the two groups of subjects on which the split-half correlation was computed, the fit is 0.89 and 0.85, respectively. The fit of the model can also be expressed by the linear trend in the RTs for the number of steps in the model (see Fig. 7.2). This linear trend is significant: $F(1,336) = 68.81$, $p < 0.001$.

7.2 Experiment 2

The criterion in the present experiment was the parent-child relation: subjects had to judge whether the two terms express a parent-child relation. Again it was assumed that the criterion directs the process: subjects will first try to verify the information with respect to the criterion. If the information cannot be verified, a falsification stage follows. Just as in the previous experiment, the structure of the material will affect the RT. Parent terms will more readily be processed than sibling terms, and these will more readily be processed than child terms. There are, however, two important qualifications with respect to this order. In previous experiments it was found that when subjects have to judge whether two concepts are both a *tree*, the RT is very short if none of the concepts is related to the criterion *tree*. This is the case, e.g., for *sparrow fly*. Accordingly, it was expected that the RT for an item where both terms are neither parent nor child is very short. The items in question are *brother sister* and *sister brother*. The second qualification relates to the criterion. The terms have to be judged with respect to the parent-child relation. Consequently, it was expected that in the verification stage sibling terms are searched for later than child terms. Furthermore, it was expected on the basis of the data of the first experiment that the sex of the subjects will interact with the items. This interaction was expected to occur at least with the *father* and *mother* items. Finally, a prediction was made with respect to the processing of male and female terms. There appears to be an asymmetry between male and female kinship terms, just as between unmarked and marked terms [F.10]. The meaning component male is more easily assumed and inferred than the meaning component female: the processing of male kinship terms is easier than the processing of female kinship terms. This effect could have occurred in the previous experiment, but was not observed. This in fact makes sense, given the criterion sex. If a subject has found the parent of his own sex, he will search for another kinship term of the same sex. Similarly, once he has found a parent of the opposite sex, he will search for a term that matches with respect to the sex.

7.2.1 Method

Material. The experimental material consisted of 16 word pairs of the kin terms *father, mother, son, daughter, brother,* and *sister*. The items can be considered as 9 items types if the left-right order is not taken into account. The items are presented in Table 7.2. Other combinations of the six kin terms served as practice items. The items were presented in 4 experimental blocks of 18 items each. These blocks were preceded by two practice blocks. There were two practice items at the beginning of each block. The word pairs were distributed equally among all parts of the blocks; the sums of the rank orders for the positions of the word pairs in the blocks were approximately equal.

The number of items for which the correct answer is "yes" was approximately the same as the number of items for which the correct answer is "no". The sequence of items was such that not more than three consecutive items required the same answer, and that no word occurred more than once on two successive items.

Procedure. The procedure was the same as in Experiment 1 of this chapter, except for the fact that the answer buttons were labeled "yes and "no", respectively.

Subjects. The subjects were 20 students, the majority of them undergraduates in psychology or biology. They were paid for their participation. One-half of them were male, the other half female.

Table 7.2. Mean RT in ms for judgments with respect to the criterion parent-child relation (Experiment 2)

Item		Sex of subjects Male	Female
father son	/son father	818	789
father daughter	/daughter father	986	852
mother daughter	/daughter mother	976	800
mother son	/son mother	909	787
father mother	/mother father	897	813
son daughter	/daughter son	1126	1021
brother sister	/sister brother	851	777
father brother		979	981
sister mother		1041	934

7.2.2 Results

The number of errors averaged (as in Experiment 1) 1.9%. No subject made more than 5% errors. For each subject 16 medians were computed for the correct responses only, one for each item. A three-way analysis of variance was performed on these medians. There were two between subjects factors: assignment of the right button to the answer "yes" or to the answer "no" and sex of the subjects. The within subjects factor was the factor items with 16 levels.

The predictions were confirmed by the data. The RT for items requiring the answer "yes" was 67 ms shorter than the RT for the items requiring the answer "no": F (1,240) = 28.89, $p < 0.001$. The RTs for the items *brother sister* and *sister brother* were nearly the shortest RTs for all the items. That the RT for parent items is shorter than for child items is indicated by the 219 ms difference between the item types *father mother* and *son daughter*: $F(1,240) = 77.01$, $p < 0.001$. The sex of the subjects interacted with the items containing a parent term: the RT was 38 ms shorter if the parent was of the same sex as the subject than when the parent term was of the opposite sex: $F(1,240) = 5.74$, $p < 0.05$. The difference in RTs between item types *father son* and *father daughter* and between item types *mother son* and *mother daughter* was on the average 78 ms, the son items being shorter than the daughter items, as was predicted on the basis of the male-female asymmetry: $F(1,240) = 19.54$, $p < 0.001$.

Some other results should be mentioned. The order of the presentation of the kinship terms had no effect on the RTs. The difference in RT between two items for all pairs of items that differed only with respect to this order did not reach significance. There were 7 cases of such item pairs. The 16 items are treated, therefore, as 9 item types. The RTs in Table 7.2 are averaged over the two items in each type, separated for males and females.

There appeared to be a small but significant effect of the semantic similarity: the RT for the "yes" answers was 38 ms shorter when both terms were of the same sex than when they were of opposite sex: $F(1,240) = 4.64$, $p < 0.05$; if one defines the related and unrelated items with respect to judged similarity data (Table 6.1), the RT for the "yes"-related items is again these 38 ms shorter than the RT for "yes"-unrelated items. However, the RT for "no"-related items is not longer but even 70 ms shorter than the RT for the "no"-unrelated items. The product moment correlation between similarity judgments and RTs for items requiring the answer "yes" was -0.43, and for items requiring the answer "no" negative: -0.51.

A split-half correlation was computed on the item types, as in the previous experiment, in order to test the reliability of the data. Subjects were divided into two groups, which were matched with respect to sex of the subjects and left-right key assignment. The average product moment correlation was 0.91 (for male subjects 0.95 and for female subjects 0.86).

7.2.3 Discussion

The results provide evidence for the following factors underlying judgment processes. Subjects attempt to find positive evidence with respect to the

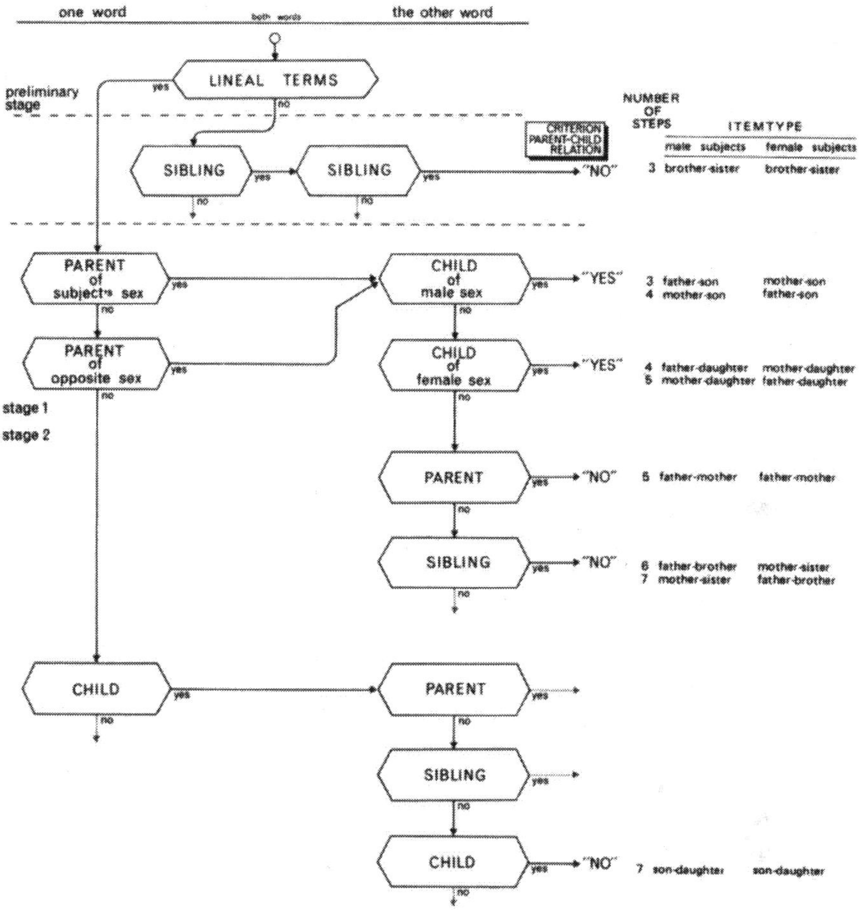

Fig. 7.3. Process model for the verification of kinship terms with respect to the criterion parent-child relation in Experiment 2

criterion; if this evidence cannot be found, a falsification process follows. There appears to be an order of searching information. First, items that do not contain any parent or child term are eliminated. Second, evidence on parent terms is searched for earlier than evidence on child terms. This fact, combined with the observed interaction of the sex of the subject with the sex of the parent term, leads to searching first for the parent term of subject's sex. This may again be interpreted as the orientation of the subject. Finally, the male-female asymmetry, combined with the search for positive information with respect to the criterion parent-child relation, implies that after the question about parents, the question about a male child is asked. The factors are the same as those in Experiment 1, except for the male-female asymmetry and for the *brother sister* item type. These differences can be interpreted in terms of the properties of the tasks.

These factors specify the model as presented in Fig. 7.3. The first step is to check whether there are any "lineal" terms. If this is not the case, the answer "no" is given when positive evidence is obtained that both terms are siblings. If there is a lineal term, the first question in the process regards the parent of the subject's sex. If this term is not found, the process continues searching for the parent of the opposite sex. If one of these questions is answered positively, questions are asked with respect to the other term: male child, female child, respectively. If the answer is "no", the falsification process starts with the standard order of the previous experiment: parent, sibling, child. If none of the questions about parent terms is answered positively, the process continues with similar questions about child terms.

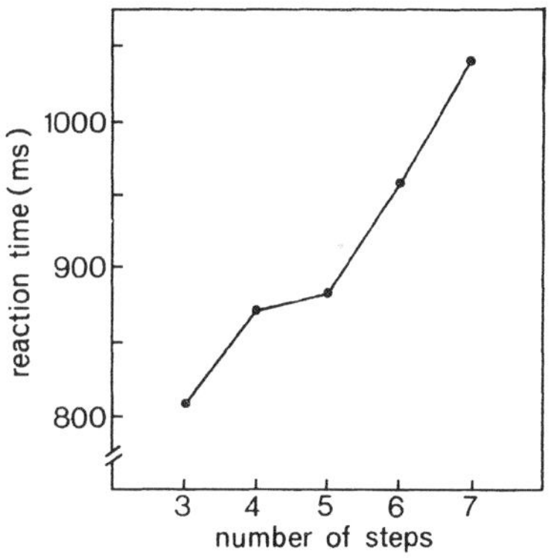

Fig. 7.4. The fit of the model to the results in Experiment 2

The fit of the model can again be expressed by the linear trend in the RTs for the items differing in number of steps according to the model (see Fig. 7.4). The number of steps varies from 3 to 7. The linear trend is significant: $F(1,240) = 177.91$, $p < 0.001$. The fit of the model in terms of a correlation between number of steps and RTs, computed with the Pearson r, is 0.94. For the two groups on which the split-half correlation was computed, the fit is 0.92 and 0.94, respectively.

One result has not yet been accounted for in the model: the effect of semantic similarity for the "yes" answers. This can be done, however, in the following rather speculative way. The transition from the question "child of male sex" to "child of female sex" for male subjects is more difficult when a male parent has already been found than when a female parent has been found. For female subjects, the same transition is easier when a female parent has been found. This point, however, deserves further study.

7.3 General Discussion and Conclusion

A few differences between the studies of the retrieval process in the hierarchical domain and in the kinship domain should be discussed. First, the difference between the two domains is reflected in the way in which the retrieval processes were described. The verification processes in the hierarchical domain were described in terms of the retrieval of categories. No attention was paid to the individual words: words within a category were simply treated as exemplars of that category. In this sense one can call the model a global model. The results of the kinship term experiments, on the other hand, were described by a detailed model: the retrieval processes are described not in terms of hierarchical categories but in terms of operations on the individual words.

Secondly, the difference between the two domains is reflected in the role of the criterion in the verification processes. The relations between the features in the hierarchical domain are set-superset relations: the features are nested. The criterion is a feature from the hierarchy that itself is related to other features in the domain in terms of set-superset relations. Consequently, one feature may be informative with respect to another feature and with respect to the criterion. This is the reason why, for example, the RT for *oak* and *beech* with respect to the criterion *plant* is shorter than the RT for *oak* and *tulip*: the similarity of the concepts in the first word pair with respect to *tree* facilitates the judgment with respect to the criterion. In the kinship domain, on the other hand, the feature that is the criterion is not related to the other features. Therefore, the role of the criterion in actualizing the structure of the domain is illustrated especially by the kinship study: the order of information to be retrieved varies with the criterion.

Another point should be discussed with respect to the role of the criterion in the two domains. It has been argued that semantic similarity between concepts in the hierarchical domain is intrinsically related to the criterion, e.g., the similarity between *oak* and *beech* as well as between *oak* and *tulip* is related to the criterion *tree*. Accordingly, the similarity between the concepts can be described to a large extent in terms of the similarity of each of the concepts with respect to the criterion. This was illustrated in Sect. 4.4 (Table 4.8). The elaboration of the feature comparison model of SMITH et al. presented by SHOBEN [1.29] is in agreement with this argument. In their model the semantic similarity between each exemplar and the target category determines the RT, while the similarity between the exemplars of the word pair does not play a role. It should be noted, however, that such a model cannot account for the results on domains in which the comparison between concept and criterion cannot be made. For example, it is not possible to determine the similarity between a kin term and sex.

What can be concluded with respect to the question that motivated the study in this part? The main principles of the model, developed for the hierarchical domain, are generalized to the kinship domain. First, the verification/falsification order is confirmed, both on judgments with respect to sex (Experiment 1) and on judgments with respect to the parent-child relation (Experiment 2). Second, factors derived from the structure of the kinship domain guide the processes in the verification and falsification stage, as is indicated by the order of processing parent, sibling, and child terms.

Chapter 8 Summary and Conclusion

8.1 Summary

The present study was concerned with the retrieval of semantic information. Retrieving semantic information is a fundamental process in almost any kind of cognitive behavior. The Introduction presented the main experimental paradigms and results found in the literature on semantic memory as well as an outline of the most important models.

The tasks that generally were used are simple verification tasks. In one type of tasks subjects are required to judge whether a relation exists between two concepts. The concepts are presented in a sentence, e.g., *a canary is a bird* or *a canary is a sparrow*. The sentences have to be judged as true or false. In another type of tasks two concepts are presented, e.g., *oak beech* or *oak sparrow*, and the subject has to judge whether both concepts are exemplars of a prespecified category, e.g., *tree*. The concepts have to be judged as same or different with respect to the prespecified categories.

The main results of such studies can be summarized as follows. The reaction time (RT) is a function of the semantic similarity between concepts. The RT for judgments on true or same items is shorter the more similar the words are, e.g., the RT for *a canary is a bird* is shorter than the RT for *a canary is an animal*. The RT for judgments on false or different items is longer the more similar the words are, e.g., the RT for *a canary is a fish* is longer than the RT for *a canary is a flower*. These phenomena are called semantic distance effects for same and different judgments, respectively. Furthermore, it is observed from the results presented in the literature that the RT for true (same) items is shorter than the RT for false (different) items, e.g., the RT for *a canary is a bird* is shorter than the RT for *a canary is a fish*. This phenomenon is termed the same/different effect.

Chapter 2 discussed the statistical analysis of the RT data, focussing specifically on the use of medians.

The main body of this study consists of three parts.

A model for verifying semantic relations. The first part dealt with a hierarchically organized domain. The task used in this part required subjects to judge whether two concepts are same or different with respect to prespecified categories. The selection of the material for these experiments was discussed in Chap. 3. The material consisted of concepts that constitute a three-level hierarchy, viz., exemplars of the categories *tree, flower, bird*, and *insect*. In this way several types of same and several types of different relations between the concepts could be obtained. In the first experiment, production frequency data on exemplars from the categories *tree, flower, bird*, and *insect* were obtained. In the three subsequent experiments of Chap. 3 it was established that the selected high-frequency exemplars did indeed form four groups in the perception of the subjects (Experiment 2); that *tree, flower, bird,* and *insect* are natural category names of the exemplars (Experiment 3); and that *plant* and *animal* are natural supercategory names of the exemplars (Experiment 4). The former four categories appear to be more salient than the latter two.

The results of the first two verification experiments of Chap. 4, in which the prespecified categories were *tree, flower, bird,* and *insect* for the first experiment and *plant* and *animal* for the second, confirmed the same/different effect and the semantic distance effects for different judgments as well as for same judgments. These effects were found not only for items in which both words were exemplars (Ex-Ex items) but also for items in which one word was an exemplar and the other one a category (Cat-Ex items). Furthermore, the RT for Cat-Ex items was shorter than the RT for Ex-Ex items, although this difference did tend to disappear very rapidly in time.

A two-stage model that accounts for the data was described. The process can be characterized by the following two principles. First, the process starts searching for positive evidence with respect to the criterion to be verified (stage 1: verification stage). If such evidence is not found, evidence must be obtained for the conclusion "different" (stage 2: falsification stage). Second, the order of information retrieval within these two stages appears to depend on the structure of the semantic domain as actualized by the task and the criterion.

The third verification experiment excluded the possibility that the same/different effect found in Experiments 1 and 2 is due to the fact that the different-unrelated words are rather related after all: the same/different effect occurs just as well when the different-unrelated words are actually far apart in meaning. The fourth experiment showed the same/different effect as well as the semantic distance effects in a slightly different condition: there is only one target at a time with respect to which the judgments are made. Furthermore, it appeared that when both words that have to be judged differ substantially in meaning from the target, a shortcut is made in the process. Finally, no evidence was obtained for a left-to-right processing of the words. The results of the experiments were discussed in terms of the model.

Generalization to another task. In Part II the question of whether the principles of the model can be generalized to a task different from the verification task was studied. In the experiment of Chap. 5 the task required judgment of semantic similarity between two words on a seven-point scale. This task does not stress the hierarchical organization of the material. A high correlation was found between the similarity judgments and the a priori classification of the item types on the basis of their hierarchical relations. The RT is shortest for the extreme similar judgments, significantly longer for the extreme different judgments, and the longest for the intermediate similarity judgments. These RTs, however, not only depend on the overall judged semantic similarity between the concepts, they are also substantially determined by the hierarchical relations in the material. The results, therefore, indicate that the principles of the model developed in Part I can be generalized to a nonverification task.

Generalization to another semantic domain. Can the principles of the model also be generalized to a different semantic domain, a field with a paradigmatic organization? To study this, two verification experiments were performed in Part III on the kinship domain with terms of the nuclear family. First, however, the organization of the domain had to be studied: in Chap. 6 the semantic similarity between kinship word pairs and the retrieval order of the kinship terms were investigated. In a triadic comparison task (Experiment 1) the semantic similarity between kinship terms of the nuclear family had to be judged. On the basis of these judgments the kinship word pairs could be partitioned in groups of semantically more similar and semantically less similar word pairs. The retrieval order of the kinship terms was investigated by requiring subjects to name the words that denote family members and relatives (Experiment 2). Both the production frequency and the mean position in the naming sequence indicate that *father* and *mother* are salient terms, immediately followed in saliency by *brother* and *sister*. *Son* and *daughter*, however, are very low in saliency for these subjects.

In the first verification experiment in Chap. 7, subjects had to judge whether two kinship terms were same or different with respect to the criterion sex. The two main principles of the model developed in Part I were confirmed: a clear same/different effect was found and the findings of the naming experiment in the kinship domain were reflected in the RTs. Furthermore, it turned out that subjects process kinship terms of their own sex earlier than terms of the opposite sex. No effect of semantic similarity, conceived of in terms of feature overlap, was found.

In the second verification study on the kinship domain, subjects had to judge whether the two terms expressed a parent-child relation or not. Again, the two main principles of the model developed for the verification of hierarchical relations were confirmed. Furthermore, if both terms have no relation to the criterion, the process is short-circuited similarly as in the corresponding case in the hierarchical domain. Moreover, the same result as found in the first kinship verification experiment turns up: subjects process parent terms of their own sex earlier than parent terms of the opposite sex. For the other terms, however, a male-female asymmetry was found: male terms are processed earlier than female terms, irrespective of the sex of the subject. Finally, the RT for "yes" answers is shorter when both terms are of the same sex than when they are of different sex; this semantic similarity effect can be attributed to the interaction of the sex of the subjects and the sex of the terms.

8.2 Conclusion

Relation to other studies. Most of the studies in semantic memory have especially paid attention to the semantic distance effects for same and for different judgments. A phenomenon that has not been given attention is the shorter RT for same judgments than for different judgments. The present research dealt extensively with this same/different effect as well. The model that was proposed accounts for all three phenomena.

Most studies in the field of semantic memory deal with verification tasks in a hierarchical domain. The present research extended this in two directions: first, it dealt with another task as well, viz., similarity judgments; second, not only hierarchically organized domains were studied but also a paradigmatic domain. For both extensions, substantial evidence was found confirming the more general validity of the principles of the model for the hierarchical domain.

On the other hand, however, the model was not worked out for some phenomena that were found in other research. One such phenomenon is that the RT for same judgments depends on the types of different items with which the same items are presented in a block. Another phenomenon is that the RT for a different judgment is relatively short if the judgment is on high-frequency contradictory concepts.

Relation to the network/set-theoretic distinction. As was indicated in the introductory chapter, network models and feature models can be translated into each other. The present model is neutral with respect to this distinction, although the operations are mainly described in the terminology of a network model. But this is not necessarry. For example, instead of "retrieve category" one may read "compute category" just as well.

A difference between current network models and set-theoretic models is that the relations between the concepts are not of equal importance according to existing network models, whereas similarity in available set-theoretic models can be defined in terms of all possible features. In the latter type of model the semantic similarity is determined by a variety of relations (features), the hierarchical being just one of them. The hierarchical relations, however, are of outstanding importance in judgments on the hierarchical domain. Theories, such as the current set-theoretic models, that do not allow for this influence of the structure of the material and the role of the criterion in actualizing this structure will be inadequate as was demonstrated in Chaps. 5 and 7: similarity between concepts defined in terms of all possible features cannot sufficiently account for the data on the similarity judgments; the results on the kinship domain demonstrate at most only a marginal interaction effect between verification judgments and the similarity in terms of semantic features.

Relation to cognitive functioning. The processing of semantic information as revealed by the present studies is characterized by two basic factors. The first may be called a set for verification. If one has to judge semantic information, there appears to be a tendency to search for evidence that justifies the answer "true". It is only when this evidence is not found that evidence is searched for which enables one to conclude "false". The second factor is the organization of the semantic domain as actualized by the task and particularly by the criterion. In the experiments on the hierarchical domain, the criterion is a feature from the hierarchy that is itself related in terms of set-superset relations to the other features of the domain. In the kinship domain, however, the feature that is the criterion is not related to the other features. Accordingly, the role of the criterion in actualizing the structure of the domain is illustrated especially by the kinship experiments.

These two basic factors characterize the process as a goal-directed process. At each step in the process one has to search for verifying or falsifying information, depending on the stage of the process. The information to be searched for at each step depends on the actualized organization of the semantic domain.

Evidence for the conclusion that subjects try first to verify information and only then try to falsify information was also reported by JOHNSON-LAIRD and WASON [8.1] in experiments in which subjects had to indicate which information they needed in order to decide whether a conditional sentence is true or false (but see also BRÉE [8.2]).

This description of the semantic processes fits in nicely with the research of BRUNER et al. [1.40] on concept attainment tasks. These authors made a distinction between direct tests and indirect tests. A direct test of a hypothesis is a test of what the hypothesis is about. The result of an indirect test requires a transformation in order to yield evidence on the hypothesis. BRUNER et al. claimed that "the most characteristic feature of cognitive activity [is that] human subjects prefer a direct test of any hypothesis they may be working on" [Ref. 1.40, p.86]. The processes in the present research may be described in terms of direct tests. The criterion with respect to which the information has to be judged specifies the hypothesis to be tested. This hypothesis is tested with a direct test. First, verifying evidence is searched for. If no such evidence is found, a falsification stage follows as a kind of double check. The double check itself can be considered as a direct test again. BRUNER et al. concluded that "we do not fully accept the possibilities of correctness and incorrectness being mutually exclusive" [Ref. 1.40, p.86]. Using their terminology the conclusion of the present research is that retrieval processes of semantic information can be characterized as the successive testing of hypotheses by means of direct tests.

References

F.1 D.E. Broadbent: Psychological aspects of short-term and long-term memory. Proc. R. Soc. Lond. B*175*, 333-350 (1970)
F.2 W. James: *The Principles of Psychology* (Macmillan, London 1890)
F.3 J.S. Bruner: "Modalities of Memory", in *The Pathology of Memory*, ed. by G.A. Talland, N.C. Waugh (Academic Press, London 1969)
F.4 E. Tulving: "Episodic and Semantic Memory", in *Organization of Memory*, ed. by E. Tulving, W. Donaldson (Academic Press, New York 1972)
F.5 L. Wittgenstein: *Philosophical Investigations* (Blackwell, Oxford 1953)
F.6 R. Sorabji: *Aristotle on Memory* (Duckworth, London 1972)
F.7 P. Carus: *Fundamental Problems* (Open Court, Chicago 1903)
F.8 J.D. Hollan: Features and semantic memory: Set-theoretic or network model? Psychol. Rev. *82*, 154-155 (1975)
F.9 J.T. Townsend: Some results concerning the identifiability of parallel and serial processes. Br. J. Math. Stat. Psychol. *25*, 168-199 (1972)
F.10 L.G.M. Noordman: *Inferring from Language*, Language and Communication, Vol.4 (Springer, Berlin, Heidelberg, New York 1979)
F.11 J.D. Fodor, J.A. Fodor, M.F. Garrett: The psychological unreality of semantic representations. Linguistic Inquiry *6*, 515-531 (1975)
F.12 W. Marslen-Wilson, A. Welsh: Processing interactions and lexical access during word recognition in continuous speech. Cognitive Psychol. *10*, 29-63 (1978)
F.13 M.D.S. Braine: On the relation between the natural logic of reasoning and standard logic. Psychol. Rev. *85*, 1-21 (1978)
F.14 F.C. Donders: Over de snelheid van psychische processen. Onderzoekingen gedaan in het Physiologisch Laboratorium der Utrechtsche Hoogeschool *2*, 92-120 (1868-1869) [English transl.: Acta Psychol. *30*, 412-431 (1969)]
F.15 W. Wundt: *Logik* (Engelmann, Leipzig 1880)
1.1 R.S. Woodworth: *Experimental Psychology* (Holt, New York 1938)
1.2 C.E. Osgood: *Method and Theory in Experimental Psychology* (Oxford University Press, New York 1953)
1.3 D. Krech, R.S. Crutchfield: *Elements of Psychology* (Knopf, New York 1958)
1.4 W. Kintsch: *Memory and Cognition* (Wiley, New York 1977)
1.5 E.E. Smith: "Theories of Semantic Memory", in *Handbook of Learning and Cognitive Processes*, Vol.6, ed. by W.K. Estes (Erlbaum, Hillsdale, N.J., in press)
1.6 W. Noordman-Vonk, L.G.M. Noordman: "Semantisch Geheugen: Het Geheugen uit de Vergetelheid", in *Psychologie in 1975*, ed by H. Gosker et al. (Tjeenk Willink, Groningen 1975)
1.7 S. Sternberg: Memory scanning: Mental processes revealed by reaction-time experiments. Am. Sci. *57*, 421-457 (1969)
1.8 G.A. Miller: A psychological method to investigate verbal concepts. J. Math. Psychol. *6*, 169-191 (1969)
1.9 C.E. Osgood: "Interpersonal Verbs and Interpersonal Behavior", in *Approaches to Language and Thought*, ed. by J. Cowan (University of Arizona Press, Tucson 1970)

1.10 W.J.M. Levelt: "Semantic Features: A Psychological Model and Its Mathematical Analysis", Heymans Bulletins HB-69-45, Groningen (1969)
1.11 L.G.M. Noordman, W.J.M. Levelt: "Noun Categorization by Noun-Verb Intersection for the Dutch Language", Heymans Bulletins HB-70-60, Groningen (1970)
1.12 S. Fillenbaum, A. Rapoport: *Structures in the Subjective Lexicon* (Academic Press, New York 1971)
1.13 A.M. Collins, M.R. Quillian: Retrieval time from semantic memory. J. Verbal Learning Verbal Behav. *8*, 240-247 (1969)
1.14 E.E. Smith, L.J. Rips, E.J. Shoben: "Semantic Memory and Psychological Semantics", in *The Psychology of Learning and Motivation*, Vol.8, ed by G.H. Bower (Academic Press, New York 1974)
1.15 A.L. Glass, K.J. Holyoak: Alternative conceptions of semantic theory. Cognition *3(4)*, 313-339 (1975)
1.16 J.R. Anderson, G.H. Bower: *Human Associative Memory* (Winston, Washington, D.C. 1973)
1.17 W. Kintsch: *The Representation of Meaning in Memory* (Erlbaum, Hillsdale, N.J. 1974)
1.18 D.A. Norman, D.E. Rumelhart: *Explorations in Cognition* (Freeman, San Francisco 1975)
1.19 E.E. Smith, E.J. Shoben, L.J. Rips: Structure and process in semantic memory: A featural model for semantic decisions. Psychol. Rev. *81*, 214-241 (1974)
1.20 A.M. Collins, E.F. Loftus: A spreading-activation theory of semantic processing. Psychol. Rev. *82*, 407-428 (1975)
1.21 L.J. Rips, E.J. Shoben, E.E. Smith: Semantic distance and the verification of semantic relations. J. Verbal Learning Verbal Behav. *12*, 1-20 (1973)
1.22 T.K. Landauer, J.L. Freedman: Information retrieval from long-term memory: Category size and recognition time. J. Verbal Learning Verbal Behav. *7*, 291-295 (1968)
1.23 A.M. Collins, M.R. Quillian: Does category size affect categorization time? J. Verbal Learning Verbal Behav. *9*, 432-438 (1970)
1.24 D.E. Meyer: On the representation and retrieval of stored semantic information. Cognitive Psychol. *1*, 242-299 (1970)
1.25 L.J. Rips: Quantification and semantic memory. Cognitive Psychol. *7*, 307-340 (1975)
1.26 B. Schaeffer, R. Wallace: Semantic similarity and the comparison of word meanings. J. Exp. Psychol. *82*, 343-346 (1969)
1.27 B. Schaeffer, R. Wallace: The comparison of word meanings. J. Exp. Psychol. *86*, 144-152 (1970)
1.28 W. Noordman-Vonk, L.G.M. Noordman: "On Strategies in Semantic Memory", paper presented at the Spring Conference of the British Experimental Psychology Society and the Netherlands Psychonomics Foundation, Amsterdam, 1973
1.29 E.J. Shoben: The verification of semantic relations in a same-different paradigm: An asymmetry in semantic memory. J. Verbal Learning Verbal Behav. *15*, 365-379 (1976)
1.30 D. Besner, M. Coltheart: Same-different judgments with words and nonwords: The differential effects of relative size. Memory Cognition *3*, 673-677 (1975)
1.31 T.K. Landauer, D.E. Meyer: Category size and semantic memory retrieval. J. Verbal Learning Verbal Behav. *11*, 539-549 (1972)
1.32 C. Conrad: Cognitive economy in semantic memory. J. Exp. Psychol. *92*, 149-154 (1972)
1.33 M.R. Quillian: "Semantic Memory", in *Semantic Information Processing*, ed. by M. Minsky (M.I.T. Press, Cambridge, Mass. 1968)
1.34 A.M. Collins, M.R. Quillian: "How to Make a Language User", in *Organization of Memory*, ed. by E. Tulving, W. Donaldson (Academic Press, New York 1972)

1.35 E.H. Rosch: Natural categories. Cognitive Psychol. *4*, 328-350 (1973)
1.36 R.S. Nickerson: Response times for "same"-"different" judgments. Perceptual Motor Skills *20*, 15-18 (1965)
1.37 R.S. Nickerson: Same-different response times: A further test of a "counter and clock" model. Acta Psychol. *35*, 112-127 (1971)
1.38 R.C. Calfee: *Human Experimental Psychology* (Holt, Rinehart and Winston, New York 1975)
1.39 A.M. Liberman, F.S. Cooper, D.P. Shankweiler, M. Studdert-Kennedy: Perception of speech code. Psychol. Rev. *74*, 431-461 (1967)
1.40 J.S. Bruner, J.J. Goodnow, G.A. Austin: *A Study of Thinking* (Wiley, New York 1956) (Science Editions, Inc., New York 1962)
1.41 G.A. Miller, P.N. Johnson-Laird: *Language and Perception* (Cambridge University Press, Cambridge, Mass. 1976)
1.42 A. Lehrer: *Semantic Fields and Lexical Structure* (North-Holland, Amsterdam 1974)
1.43 H.H. Clark, E.V. Clark: *Psychology and Language: An Introduction to Psycholinguistics* (Harcourt Brace Jovanovich, New York 1977)
2.1 H. Scheffé: *The Analysis of Variance* (Wiley, New York 1959)
2.2 W.L. Hays: *Statistics for Psychologists* (Holt, Rinehart and Winston, New York 1963)
2.3 J.E. Keith Smith: Data transformations in analysis of variance. J. Verbal Learning Verbal Behav. *15*, 339-346 (1976)
2.4 D.H. Krantz, A. Tversky: Conjoint-measurement analysis of composition rules in psychology. Psychol. Rev. *78*, 151-169 (1971)
2.5 D.H. Krantz, R.D. Luce, P. Suppes, A. Tversky: *Foundations of Measurement*, Vol.1 (Academic Press, New York 1971)
2.6 H.H. Clark: The language-as-fixed-effect fallacy: A critique of language statistics in psychological research. J. Verbal Learning Verbal Behav. *12*, 335-359 (1973)
2.7 B.J. Winer: *Statistical Principles in Experimental Design* (McGraw-Hill, New York 1971)
2.8 K.I. Forster, R.G. Dickinson: More on the language-as-fixed-effect fallacy: Monte Carlo estimates of error rates for F_1, F_2, F' and $min\ F'$. J. Verbal Learning Verbal Behav. *15*, 135-142 (1976)
2.9 E.L. Wike, J.D. Church: Comments on Clark's "The language-as-fixed-effect fallacy". J. Verbal Learning Verbal Behav. *15*, 249-255 (1976)
2.10 H.H. Clark, J. Cohen, J.E. Keith Smith, G. Keppel: Discussion of Wike and Church's comments. J. Verbal Learning Verbal Behav. *15*, 257-266 (1976)
2.11 R.E. Kirk: *Experimental Design: Procedures for the Behavioral Sciences* (Brooks/Cole, Belmont, Calif. 1968)
3.1 J. Linschoten: "De la Court's frekwentietelling van Nederlandse woorden", Psychologisch Laboratorium Rapport No. 6301, Utrecht (1963)
3.2 P.C. Uit den Boogaart (ed.): *Woordfrequenties in geschreven en gesproken Nederlands* (Oosthoek, Scheltema en Holkema, Utrecht 1975)
3.3 W.F. Battig, W.E. Montague: Category norms for verbal items in 56 categories: A replication and extension of the Connecticut category norms. J. Exp. Psychol. Monogr. *80(2)*, 1-46 (1969)
3.4 E.F. Loftus, R.W. Scheff: Categorization norms for fifty representative instances. J. Exp. Psychol. Monogr. *91*, 355-365 (1971)
3.5 W. Noordman-Vonk: "Categorie normen voor woorden in 34 categorieën", Heymans Bulletins HB 78-373, Groningen (1978)
3.6 G.A. Miller: "Empirical Methods in the Study of Semantics", in *Journeys in Science: Small Steps - Great Strides*, ed. by D.L. Arm (University of New Mexico Press, Albuquerque 1967)
3.7 S.C. Johnson: Hierarchical clustering schemes. Psychometrika *32*, 241-254 (1967)

4.1 P.A. Carpenter, M.A. Just: Sentence comprehension: A psycholinguistic processing model of verification. Psychol. Rev. *82*, 45-73 (1975)
5.1 W. Noordman-Vonk: "Response Latency and the Discrimination of Response Button Position", in preparation
5.2 A. Caramazza, H. Hersh, W.S. Torgerson: Subjective structures and operations in semantic memory. J. Verbal Learning Verbal Behav. *15*, 103-117 (1976)
5.3 W. Noordman-Vonk: "Are Response Execution Time and Stimulus Evaluation Independent?", in preparation
5.4 S. Siegel: *Nonparametric Statistics for the Behavioral Sciences* (McGraw-Hill, New York 1956)
6.1 A.F.C. Wallace, J. Atkins: The meaning of kinship terms. Am. Anthropol. *62*, 58-79 (1960)
6.2 A.K. Romney, R.G. D'Andrade: Cognitive aspects of English kin terms. Am. Anthropol. *66(2)*, 146-170 (1964)
6.3 M. Bierwisch: "Semantics", in *New Horizons in Linguistics*, ed. by J. Lyons (Penguin, Harmondsworth, Baltimore 1970)
6.4 K.N. Wechsler, A.K. Romney: "Individual Variations in Cognitive Structures", in *Multidimensional Scaling*, Vol.2, ed. by A.K. Romney, R.N. Shepard, S.B. Nerlove (Seminar Press, New York 1972)
6.5 S.E. Haviland, E.V. Clark: "This man's father is my father's son": A study of the acquisition of English kin terms. J.Child Language *1*, 23-47 (1974)
6.6 D.I. Slobin: *Psycholinguistics* (Scott Foresman, Glenview, Ill. 1971)
8.1 P.N. Johnson-Laird, P.C. Wason: A theoretical analysis of insight into a reasoning task. Cognitive Psychol. *1*, 134-148 (1970)
8.2 D.S. Brée: "The Interpretation of Implication", in *Artificial and Human Thinking*, ed. by A. Elithorn, D. Jones (Elsevier, Amsterdam 1973)

Author Index

Anderson, J.R. [1.16] 3
Atkins, J. [6.1] 63
Austin, G.A. [1.40] 8,88

Battig, W.F. [3.3] 18
Besner, D. [1.30] 4
Bierwisch, M. [6.3] 63
Bower, G.H. [1.16] 3
Braine, M.D.S. [F.13] IX
Brêe, D.S. [8.2] 88
Broadbent, D.E. [F.1] V,1,3
Bruner, J.S. [F.3,1.40] V,VI,8,88

Calfee, R.C. [1.38] 7
Caramazza, A. [5.2] 52
Carpenter, P.A. [4.1] 34
Carus, P. [F.7] VII,IX
Church, J.D. [2.9] 13
Clark, E.V. [1.43,6.5] 8,63
Clark, H.H. [1.43,2.6,2.10] 8,12, 13
Cohen, J. [2.10] 13
Collins, A.M. [1.13,1.20,1.23,1.34] 2,3,4-5,33,34
Coltheart, M. [1.30] 4
Conrad, C. [1.32] 5
Cooper, F.S. [1.39] 8
Crutchfield, R.S. [1.3] 1

D'Andrade, R.G. [6.2] 63,64,65,68
Dickinson, R.G. [2.8] 13
Donders, F.C. [F.14] IX,2

Fillenbaum, S. [1.12] 2,63,64
Fodor, J.A. [F.11] IX
Fodor, J.D. [F.11] IX
Forster, K.I. [2.8] 13
Freedman, J.L. [1.22] 3

Garrett, M.F. [F.11] IX
Glass, A.L. [1.15] 2,5,10
Goodnow, J.J. [1.40] 8,88

Haviland, S.E. [6.5] 63
Hays, W.L. [2.2] 11
Hersh, H. [5.2] 52
Hollan, J.D. [F.8] VII,7
Holyoak, K.J. [1.15] 2,5,10

James, W. [F.2] V
Johnson, S.C. [3.7] 21
Johnson-Laird, P.N. [1.41,8.1] 8, 88
Just, M.A. [4.1] 34

Keith Smith, J.E. [2.3,2.10] 11,12, 13
Keppel, G. [2.10] 13
Kintsch, W. [1.4,1.17] 1,3,7
Kirk, R.E. [2.11] 13
Krantz, D.H. [2.4,2.5] 11
Krech, D. [1.3] 1

Landauer, T.K. [1.22,1.31] 3,4
Lehrer, A. [1.42] 8

Levelt, W.J.M. [1.10,1.11] 2
Liberman, A.M. [1.39] 8
Linschoten, J. [3.1] 18
Loftus, E.F. [1.20,3.4] 3,5,18,34
Luce, R.D. [2.5] 11

Marslen-Wilson, W. [F.12] IX
Meyer, D.E. [1.24,1.31] 3,4
Miller, G.A. [1.8,1.41,3.6] 2,8,21
Montague, W.E. [3.3] 18

Nickerson, R.S. [1.36,1.37] 7
Noordman, L.G.M. [F.10,1.6,1.11, 1.28] IX,2,3,76
Noordman-Vonk, W. [1.6,1.28,3.5, 5.1,5.3] 2,3,19,52
Norman, D.A. [1.18] 3

Osgood, C.E. [1.2,1.9] 1,2,8

Quillian, M.R. [1.13,1.23,1.33, 1.34] 2,3,4,5,33,34

Rapoport, A. [1.12] 2,63,64
Rips, L.J. [1.14,1.19,1.21,1.25] 2,3,4,6,8,82
Romney, A.K. [6.2,6.4] 63,64,65,68
Rosch, E.H. [1.35] 6
Rumelhart, D.E. [1.18] 3

Schaeffer, B. [1.26,1.27] 3,4,6, 30,39
Scheff, R.W. [3.4] 18
Scheffé, H. [2.1] 11
Shankweiler, D.P. [1.39] 8
Shoben, E.J. [1.14,1.19,1.21,1.29] 2,3,4,6-7,8,43,82
Siegel, S. [5.4] 55
Slobin, D.I. [6.6] 68
Smith, E.E. [1.5,1.14,1.19,1.21] 1,2,3,4,6-7,8,82
Sorabji, R. [F.6] VII
Sternberg, S. [1.7] 2
Studdert-Kennedy, M. [1.39] 8
Suppes, P. [2.5] 11

Torgerson, W.S. [5.2] 52
Townsend, J.T. [F.9] IX
Tulving, E. [F.4] V,1,3
Tversky, A. [2.4,2.5] 11

Uit den Boogaart, P.C. [3.2] 18

Wallace, A.F.C. [6.1] 63
Wallace, R. [1.26,1.27] 3,4,6,30,39
Wason, P.C. [8.1] 88
Wechsler, K.N. [6.4] 63
Welsh, A. [F.12] IX
Wike, E.L. [2.9] 13
Winer, B.J. [2.7] 13,14
Wittgenstein, L. [F.5] VI
Woodworth, R.S. [1.1] 1
Wundt, W. [F.15] IX

Subject Index

Actualization, of information 2
Analysis of variance 10-14
additive model 12,13
Geisser-Greenhouse procedure 14
homogeneity of variances 11,12
least square solution 10
linear model 12
logarithmic transformation 11-12
min F' 13
missing data 10,11,12
multiplicative model 12
nonlinear transformation 11
unweighted means analysis 11,12
use of the median 12,13,83
variance-covariance matrices 13-14
weighted means analysis 10-11

Categorical perception, of speech sounds 8
Category 17-18
and exemplar 3,17, *see also* Word pair
degree of membership in 6
higher order 17,21,26,27-28
norms 18,19
prespecified 3,6,17, *see also* Category, target
subcategory 17,24
supercategory 23-24,25,26,27-28
target category 17,25,29,32,33,36 36,41,42, *see also* Criterion

Componential system, of kinship domain 63-64
Components, in kinship domain 63, 65
Criteriality, of paths, *see* Network model
Criterion, in verification task 64, 69, *see also* Category, target
role of, 64,68,81-82,87
sex 63,70-75,85
parent/child relation 63,75-80, 86

Decision strategy, hierarchical 33
Double check, in verification/falsification process 34,88

Episodic memory, and semantic memory 1
Error rate, correlation with RT 10
Exemplar, and category 3,29, *see also* Word pair
Exemplars, categorization of 23-24
categorization of pairs of 26-28
grouping of 21-23
production of 18-20,39-40

Falsification stage 33,34,47,84, *see also* Verification/falsification order
Feature comparison model 6-7, *see also* Set-theoretic model

Features 2,4,6
defining and characteristic 6
overlap of sets of 6,7,69
Feature set model 6,39
Frequency count, of words 18

Goal directedness, of process 48,87

Hierarchically organized domain 3, 8,9,17-28,81,84
Hierarchical organization 8,47,51 58,59-60,81,87
three-level 17,24,26,28,84

Information processing approach 2

Kinship, naming sequence of terms 66-68,85
parent, sibling, child terms 68,72,73,76,81
Kinship domain 8,63-68,81,87

Lexicon, subjective organization of 2,20

Male-female asymmetry 76,78,79,86
Marker search model 5-6
contradictory sentences 6,87
Matching process 33-34,47

Network model 4-6
accessibility of paths 5
and set-theoretical model 7,87
criteriality of paths 5
nodes and paths in 4,5
spreading of activation 34

Organization of semantic domain, actualized 47,51,63,64,69,81,82, 85,87
subjective 17, *see also* Lexicon

Paradigmatic field 8,9,63,85
Personal orientation 66-68,73,79
Preliminary stage 47,79-80
Processing order, left to right 43, 44-46,48,72,74-75,78,85
Production frequency, of exemplars 19,39-40,84
of kinship terms 66-67,85

Reaction time 2,10
distribution of 11,13
median 12,13,83
Relation, between exemplar and target 3,43-44,46-47
between exemplars 3,24, *see also* Word pair
of inclusion 3, *see also* Hierarchical organization
set-superset 81, *see also* Hierarchical organization
Response button position 52
Response execution, button panel vs voice key 52-53,54-55
Retrieval order, of kinship terms 63-64,85

Saliency, of categories 26,53,84
of kinship terms 64,66,68,85
Same/different effect 4,7,8,29,30, 33,36,38,39-40,42,46,57,60,70,72 78,83,84,86
Same/different judgment 3,7
Same sex effect 72,78,79
Sampling model 39, *see also* Set-theoretic model
Sampling process, in set-theoretic model 4,6
Search, time limited 34
Semantic components, *see* Components

Semantic distance 4-5
effect 4,6,7,8,36,70,83,84,86
effect for different word pairs 4,29,30,33,42,46,83
effect for same word pairs 4,29,36,38,83
Semantic features, *see* Features
Semantic field, 8, *see also* Hierarchical organization *and* Paradigmatic field
Semantic memory 1-2
acquisition of 2
and episodic memory 1
large-scale models 2-3
local models 2
Semantic relation, *see* Relation
Semantic similarity 4,51,65,69,85,87, *see also* Relation
between exemplar and criterion 70
between exemplars 51-60,85
between kinship terms 65-66,69,85
effect 4,70,73,78,81,86, *see also* Semantic distance, effect
rated 51-60,64,66,70,85
Sentence verification task 3,6
Set-theoretic model 4,6-7,39,87
thresholds in 4,7,8,39
Similarity judgment 51-60,63,65-66,85
Spreading of activation 34

Target category, *see* Category, target
Triadic comparison 65
True/false judgment 3
Truth index 34

Verification, set for 63,64,68,87, *see also* Verification/falsification order
Verification/falsification order 33,47,51,66,69,73,76,79-80,82,84,88
Verification stage 33,34,47,84, *see also* Verification/falsification order

Word comparison task 3,6
Word pair, Cat-Ex 29-30,32-33,36-38,84
different 25,36-38,44,46,84
different-related and different-unrelated 25-26,26-27,29-30,32-33,40-42,44,46,53,56,58-60,73,78
Ex-Ex 29-30,32-33,36-38,84
same 25,26-27,29-30,32-33,40-42,44,46,53,56,58-60,84
same-related and same-unrelated 25,36-38,73,78

H. Hörmann

Psycholinguistics

An Introduction to Research and Theory
Translated from the German edition by
H. H. Stern
1971. 69 figures. XII, 377 pages
ISBN 3-540-05159-7

„...provides a comprehensive introduction to the psychology of language by concentrating on the behaviourist conception...
the translation is written in a clear, concise and compact English...
The substance of this book, which has become a standard textbook in German as well as the brilliancy of its translation will certainly secure its position in the English speaking world as well."

IRAL (Deutschland)

B. Malmberg

Structural Linguistics and Human Communication

An Introduction into the Mechanisms of Language and the Methodology of Linguistics
(Reprint of the 2nd revised edition 1967)
1976. 88 figures. VIII, 213 pages
(Kommunikation und Kybernetik in Einzeldarstellungen, Band 2)
ISBN 3-540-03888-4

Contents:

Introduction. – Signs and Symbols. The Linguistic Sign. – The Communic Process. – Preliminary Expression Analysis. Acoustic and Physiological Variable Information. – Segmentation. Forms of Expression. Oppositions and Distinctions. – Paradigmatic Structures. – Redundancy and Relevancy. Levels of Abstraction. The Distinctive Feature Concept. The Binary Choice. – Syntagmatic Structures. Distribution and Probability. – Content Analysis. – The Functions of Language. – Perception and Linguistic Interpretation. – Primitive Structures and Defective Language. – Linguistic Change. – Bibliographical Notes. – Author Index. – Subject Index.

"A general survey of modern structural linguistics by B. Malmberg...
The book is essentially intended for the advanced student, but others will also find it useful, since the author manages to deal lucidly and intelligibly with a difficult subject."

The Years Work in English Studies

G. Herdan

The Advanced Theory of Language as Choice and Chance

1966. 30 figures. XVI, 459 pages
(Kommunikation und Kybernetik in Einzeldarstellungen, Band 4)
ISBN 3-540-03584-2

Contents:

Introduction.

Language as Chance I – Statistical Linguistics: Stability of Linguistic Distributions. Explanation of Stability of Linguistic Distribution. Application of the Theory of Stability of Alphabetic Distributions to a Problem of Language Mixture.

Language as Choice I – Stylostatistics: Style as a Statistical Concept. Word Count Mathematics. Style Relationships – Bi-Variate Stylostatistics. A Guide to Stylo-statistical Investigations.

Language as Chance II – Optimal Systems of Language Structure: Combinatorics on the Phonemic (Alphabetic) Level: The Combinatorial Structure of Words. Optimality of the Word-Length Distribution. Combinatorics applied to Problems of Classical Poetry. – Combinatorics on the Lexicon Level: Random Partitioning of Vocabulary – Vocabulary Connectivity. The Generalised Random Partitioning Function and Stylostatistics. The "New Statistics" on the Vocabulary Level. – Information Theory: Principles of Information Theory. Information-Theoretical Analysis as a Tool of Linguistic Research. – Language Translations as Bi-Variate Distributions of Coding Symbols.

Language as Choice II – Linguistic Duality: The Four-fold Root of Linguistic Duality. Duality as Correcting Factor-Inadequacy of Truly Semiologic Codes. Duality and Language Translation.

Statistics for the Language Seminary: Statistics of Language in the Mass: Descriptive Statistics. Statistical Inference – The Binomial Case. Statistical Inference in the Case of Multiple Classification of Events. Theory of Correlation. – Statistics of Language in the Line: The Dimension of Time in Language Statistics. Linguistic Duality and Parity.

A Survey of Past and Present-day Statistical Linguists

Bibliography. Author-Index. Subject-Index.

Made in the USA
Las Vegas, NV
10 November 2024

11366357R00066